R.F. Walter
Dr. J.L. Savage, Frank A. Banks
J.H. Miner, Bert A. Hall, A.F. Darland
O.G. Patch, C.M. Cole

THE MIGHTIEST OF THEM ALL
Memories of Grand Coulee Dam

THE MIGHTIEST OF THEM ALL:
Memories of Grand Coulee Dam

Revised Edition

L. Vaughn Downs

Published by
ASCE Press
American Society of Civil Engineers
345 East 47th Street
New York, New York 10017-2398

ABSTRACT

The Mightiest of Them All: Memories of Grand Coulee Dam (revised edition), presents the experiences of L. Vaughn Downs from the time he started working on the dam when it was in the design stage, through the construction period and into many years of actual dam operation and maintenance. He provides glimpses into the personalities connected with the project and explains the many techniques and pieces of equipment that were developed or improved as the dam was built. Mr. Downs also devotes considerable attention to problems they encountered and the solutions developed in the hope that others will learn from these situations. This revised edition brings the story up to the current period with an examination of the upkeep and condition of the dam after 50 years, and its prospects for the future.

Library of Congress Cataloging-in-Publication Data

Downs, L. Vaughn, 1908-
The mightiest of them all: memories of Grand Coulee Dam/L. Vaughn Downs.—Rev. ed.
p.cm.
Includes bibliographical references (p.) and index.
ISBN 0-87262-935-X
1.Grand Coulee Dam (Wash.)—Design and construction. 2.Grand Coulee Dam (Wash.)—History. I.Title.
TC577.W22G734 1993
627'.82—dc20   93-37713
                                                         CIP

Cover photo: Governor Clarence Martin came over from Olympia to dump the official "first concrete" in Grand Coulee Dam, December 6, 1935.

TABLE OF CONTENTS

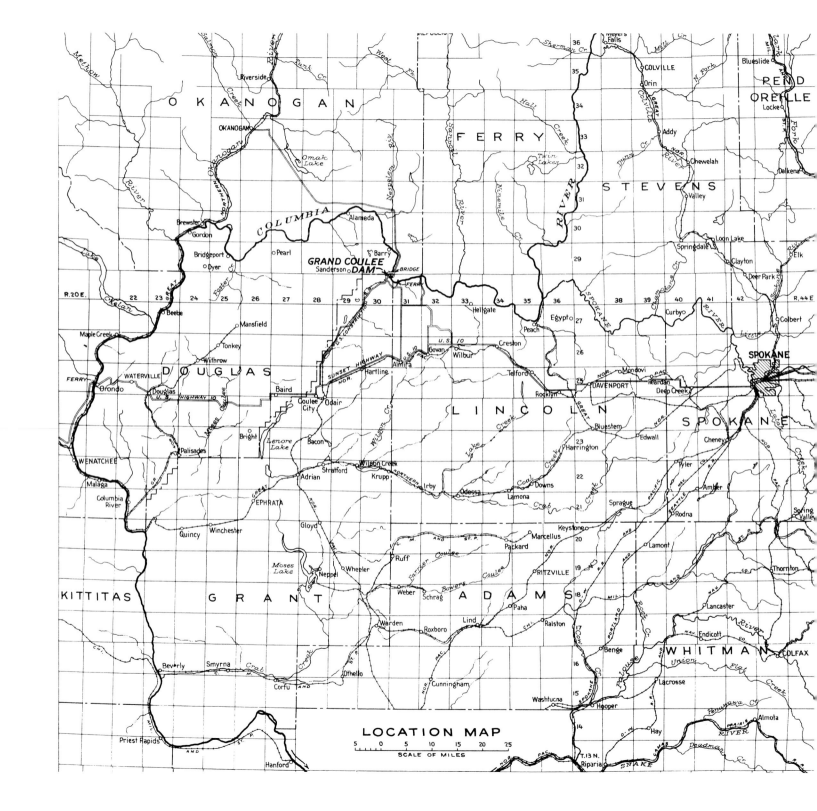

LOCATION MAP

5   0   5   10   15   20   25
SCALE OF MILES

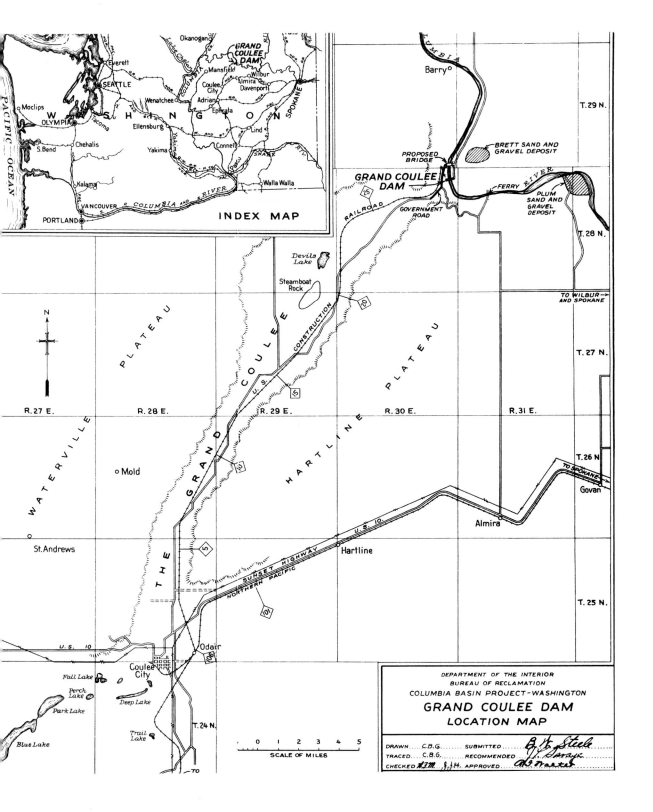

INDEX MAP

DEPARTMENT OF THE INTERIOR
BUREAU OF RECLAMATION
COLUMBIA BASIN PROJECT - WASHINGTON

GRAND COULEE DAM
LOCATION MAP

DRAWN C.B.G.	SUBMITTED	B. H. Steele
TRACED C.B.G.	RECOMMENDED	J. W. Banks
CHECKED H.J.M. J. J. H.	APPROVED	A. J. Nasten

0 1 2 3 4 5
SCALE OF MILES

An attentive crowd gave President and Mrs. Roosevelt a warm welcome when they visited the damsite August 4, 1934.

LIST OF ILLUSTRATIONS

PHOTOGRAPHS AND DRAWINGS

The photographs and drawings used in this book were all—except as noted—from the Department of the Interior, Bureau of Reclamation Official Records and have been made available to the author for inclusion herein. Most of the photographs were obtained from the files of the Grant County Historical Museum in Ephrata, Washington, the recipient of discarded Bureau photographs from the Columbia Basin Project.

Bureau photographers over the past 53 years whose photos were included were: K.S. Brown, F.B. Pomeroy, W. Fuller, S.E. Hutton, G. Hertzog, H.C. Robinson, G. Walker, R. York and B.S. Holmes.

Names in the captions are listed from left to right. I regret that photos of engineers and inspectors that supervised the inspection "on the job around the clock" have been destroyed.

Note: This is *not* an official publication of the U.S. Bureau of Reclamation. The Bureau has not sponsored this writing in any way and the inclusion of the photographs and the few drawings should not be so construed. The captions are my own elaborations.

Grand Coulee Damsite about six months after the first dirt was turned. Town of Coulee Damsite preparation at right center. Beyond lies the

Addison-Miller camp, the author's home in the fall of 1934. Cable way for construction of the bridge piers in foreground. May 1934.

Dr. John L. Savage, Chief Designing Engineer, U.S.B.R., world-famous, recipient of many honors. Born December 25, 1879, died December 28, 1967. 1952.

Frank A. Banks, Construction Engineer, Grand Coulee Dam. Born December 4, 1883, died December 14, 1957. 1945.

PREFACE

In mid-June 1985 the Kiwanis Club of Ephrata, Washington asked that I write my recollections of the construction of Grand Coulee Dam as an item to be placed in a "Time Capsule" to be installed in a niche in the Grant County Justice Center. The dedication was to be on July 27. Responding to that request—and to that of other friends over the years—I prepared a narrative in manuscript form for the capsule and the dedication was held finally on September 27, 1985. I permitted a few friends to read a copy of the narrative and I was encouraged to publish the material because of its presumed historic value.

My experience on Grand Coulee Dam was unique in that I was the only person who had been employed directly on the dam in its design stage in the Chief Engineer's Office, throughout the construction and involved as Assistant Project Manager for many years of operation and maintenance. So, on June 6, 1986 I decided to amplify the original material by adding more glimpses of certain of the personalities in the "cast" and to add many photographs and a few drawings to make the book more vivid, particularly to the families of the thousands of men and women who helped to build the dam.

Many techniques and pieces of equipment were developed or improved as the dam was built, and I mention some of them in this book to make the reader aware of the various changes—advances—that have been made in 50 years. I'm sure they will be of interest now and even more so when that Capsule is opened in the year 2084. The "unexpected happening" and the risk of such in engineering endeavors has been a continuing concern of mine. Here I have devoted considerable attention to the things that didn't work out just as designed or expected, giving emphasis to the solutions or how the problem was solved or corrected.

I have read the remarks of Mr. Edgar F. Kaiser, then president of the "Beavers" (an organization of the leaders of the heavy construction industry) at its sixth annual (1961) awards dinner honoring the outstanding builders in the United States. He was reminiscing, "If you look back; I'm sure each of you do, as I do, you look at what was accomplished and you forget the troubles." But as I look back I feel obligated to identify the *Troubles* so that similar problems may not be repeated in the future. I was told that the disaster of the Tacoma

Narrows Bridge was not the first from the same basic cause, since three or four other bridges similarly failed in the 1860s. And there have been other *Troubles* that have been experienced recurrently in succeeding generations. I would hope that this work may forestall some similar problems in the future. Hopefully it may stimulate thinking instead of second-guessing.

While I wrote most of the material for this narrative from memory, I spent considerable time reviewing project records, specifications, photographs and drawings to obtain or to verify the dates, measurements and elevations used.

In that connection I wish to acknowledge, with many thanks, my appreciation to Mr. Jim Cole, Project Manager of the Columbia Basin Project for granting permission for the inclusion of the official photographs and drawings herein. Thanks, too, to Messrs. Sid Saunders and Fay Eaton of the Engineering and Resources Division of the Grand Coulee Dam Project for their assistance in my review of official records of the dam. Mr. Saunders and his staff were most helpful in briefing me on the recent status of the Reservoir and the Spillway Bucket. I wish to acknowledge the kindness of Mr. Raymond K. Seely for his preparation of the brief supplement beginning on page 159 on some of the significant developments in his area of particular expertise and responsibility. Thanks are also expressed to Mr. John Vertrees and Mr. Borden Wilbor for the opportunity to refresh my memory with discussions on the manufacture of concrete and certain mechanical problems of their respective specialties. These three engineers, also retired from the Bureau, have been living at Coulee Dam in the very shadow—or light perhaps—of their greatest endeavor. My thanks also to Mrs. Gene Sell for taking my draft and putting it into manuscript form for that Time Capsule back in June and July of 1985.

Thanks too, to Phil Parker for his helpful suggestions and advice while I struggled in the final throes of this "book making."

Readers will note that the role of the "inspector" and quality control is emphasized. Of course that was intentional! It is not intended to detract from the work of the many brilliant engineers and builders who conceived, designed and shaped the dam, but now, when automobile manufacturers are calling back their cars by the millions to correct unsafe or defective parts, it is very comforting to look back at Grand Coulee Dam with satisfaction and the knowledge that there are few "faulty parts" buried in its great mass. The revenue from the sale

13

of power output of the Left and Right Powerplants have long since exceeded the cost of the dam and those power plants. And we, the designers, the contractors, the engineers, the inspectors and those from the ranks of crafts and labor—built it to be there for a thousand years, still a benefit to man!

As President Roosevelt closed his address at Grand Coulee on August 4, 1934, he said, "I leave here today with the feeling that this work is well undertaken; that we are going ahead with a useful project, and we are going to see it through for the benefit of our country." How prescient he was.

This is the way it was in my mind's eye. I began this effort with considerable apprehension and I finished it with a sigh of relief and a sense of pride in recounting this part of my life spent on *The Mightiest of Them All*. This is my final report on the construction of Grand Coulee Dam.

To my wife Margaret, my thanks for keeping things running (and this clean) while I got this off my chest.

L. Vaughn Downs
Ephrata, Washington
July 31, 1986

PREFACE TO THIS SECOND EDITION

This edition affords an opportunity to correct a few errors found in the first and to identify some individuals not named therein. A new chapter, Chapter XVI, examines the Low Dam/High Dam question, the upkeep and condition of the dam after 50 years, and the prospects for the future. It also features two additional photos. One identifies the personnel of the Bureau of Reclamation Chief Engineer's Office in Denver in 1931 or 1932. The other shows the widely publicized "Ice Dam." It clarifies beyond words how innovative that structure was.

My greatest reward from the book—an extra dividend, as it were—was receiving the many letters from "old timers" of all ages expressing their joy in reading the book and studying the pictures. "Thanks for doing it and thanks for the memories!" was their response—it was heart-warming.

The Author.

14

Aerial view looking southwesterly at Grand Coulee Dam and appurtenances after completion of the Third Power Plant building. Banks Lake (Equalizing Reservoir) in the distance. Upon completion of the forebay dam for the Third Power Plant as an angular extension of Grand Coulee Dam its (axis) length was increased to 5223 feet (from 4300). The volume of concrete in the structure is now 11,975,520 cubic yards. And this entire complex required 38,574,500 cubic yards of common excavation and 7,062,600 cubic yards of rock excavation. The total generating capacity is now 6,494,000 kilowatts. June 27, 1978.

When I was considering the prospect of the forthcoming civil service examination in February 1931 with Professor W.C. McNown the head of the Civil Engineering Department of the University of Kansas, he mentioned that some engineers had established good reputations in the federal service. He then mentioned his classmate, Mr. John Savage, (Wisconsin University 1903) who had a great reputation for his design work for the Bureau of Reclamation in Denver, Colorado. Professor McNown was one of the references I had used in my application. After I had been on the job in Denver about six weeks, Mr. Savage came to my drafting table to see some of the results I was getting on my assignment, so I became acquainted with this eminent engineer early in my career.

When the Bureau held an open house to display its new 5,000,000-pound capacity testing machine in the breaking of a mass concrete test cylinder 36 inches in diameter by 72 inches in height, many of the Bureau personnel and their friends and families were present. Such a test had never been performed before. There with my friends Tommie and Vickie Thomas was a young woman. "I spied you among the spectators," she recalled, "and asked, 'Who is that?'" So we were introduced. She was Margaret Savage going to art school in Denver and residing with her Uncle Jack. We were married in February 1935. So my bride and I moved into one of the new homes in Coulee Dam. Our son Richard Vaughn Downs was born in the Mason City Hospital.

I had sought an interview with Mr. Frank A. Banks, the Construction Engineer for Grand Coulee Dam when he was in Denver in the fall of 1933. He thought he could use me in his organization—just being put together.

When I arrived in Almira I sensed that I was looked upon as the "new kid on the block"—and ripe for testing. Mr. Banks, Mr. Miner, Harry Wilbert perhaps, and I were the only foreigners—non-Washingtonians—on the job. But I tried to be accommodating. When I was temporarily assigned to a survey crew doing some stadia measurements I tried to be helpful when the chief and the instrument man were reducing the notes with longhand computations. The formula required determination of the square of the cosine of the vertical angle. Having the six-place tables of natural trigonometric functions at hand, the method used was by multiplying the tabular value by itself to get the square of the value. Since I was well acquainted with trigonometry I suggested that there was a better way. When I was handed the table of values I looked at the value of the cosine of *twice* the angle, added 1 and mentally divided by 2 and read out the value sought. When their computation checked with my answer, I got only blank stares. I repeated the process for three other angles while the longhand multiplication continued, with complete disregard for my recitation of the simpler formula. I never did become aware the longhand method was abandoned—it certainly was not during the remainder of the week I worked with them.

When Grant Gordon suggested that one of the inspectors show me what the test pit crews were doing, the two of us rode up to the top of Brett pit where samples of the deposit were being taken for shipment to Denver for analysis and tests. The four-foot-square pits were being dug by hand with the walls supported by interlocking timbers which were placed as the pit deepened. My guide suggested that I ought to look at the material *in place*, so I put my feet in the bucket, grasped the ½-inch cable and was lowered 400 feet to join Hawley—the pit man—at the bottom. At the surface again, we looked over the samples and the site, then returned toward the field office. But as we got to the bottom of the hill my guide "just remembered" the rest of the crew digging test trenches at the top of the slope. Instead of driving back up there he invited me to join him as we climbed back up that steep slope over 500 feet in elevation to look at the "material in place." There were a few other "tests" of my reactions in those early days on the job. As to my guide, though, I thereafter always viewed his judgment with suspicion.

The author at the old field office at Grand Coulee Dam. ca 1936.

Raymond K. Seely, author of Chapter XIV. August 1974.

ABOUT THE AUTHOR
L. VAUGHN DOWNS

The author of this narrative, Leonard Vaughn Downs, a native of Hutchinson, Kansas, was born September 30, 1908 and attended the elementary and high schools of Inman, Kansas, where his family had moved in 1913. He attended the University of Kansas at Lawrence, Kansas, graduating with a degree of Bachelor of Science in Civil Engineering in 1931. On September 10, 1931 he entered the employment of the U.S. Bureau of Reclamation as a junior engineer on the staff of the chief engineer of that agency in its Denver, Colorado office. There his work was principally the analysis of stresses in dams, existing and proposed, using the then-new method of "trial load analysis." His assignments principally were on Boulder/Hoover Dam, the great gravity-arch structure on the Colorado River and on Grand Coulee Dam on the Columbia River. The studies on Grand Coulee Dam encompassed designs for a low dam and a high dam and on problems relating to the raising of a low dam to its ultimate dimensions. These designs typically were of straight gravity types, although preliminary designs also were analyzed for multiple arch structures. On May 18, 1934 he left Denver for Almira, Washington, the then-headquarters for the U.S.B.R. for the construction of Grand Coulee Dam to which he requested transfer as an assistant engineer. His work was in quality control—as an inspector of the contractors' work as it was being performed. His duties necessitated his availability on the site, so in September 1934 he moved into a tarpapered shack camp down on the Columbia Riverbank and in December moved into one of the bachelor quarters then constructed by the principal contractor and in the following February into the Government Camp at the base of the damsite. He resided on the site throughout the construction of the Grand Coulee Dam and related works until April 2, 1950 when he was transferred, soon to be the supervising engineer of the Ephrata, Washington office of the Columbia Basin Project. During those 16 years at the Grand Coulee Dam his work was in the inspection or supervision of inspection and acceptance of civil engineering work on excavation, backfill and cleanup in preparation for concrete,

installation of reinforcing steel and other embedded items, placement of concrete, drilling and testing of the foundation rock, foundation grouting, repair of defective concrete, observing and reporting on damage to gates and metalwork, finishes of concrete surfaces, painting of embedded metalwork and of machinery, and also served as an observer on the contractors' effort in the care and diversion of the river, the construction and removal of cofferdams, reviewed contractors' claims for adjustment and served as liaison to the contractors in resolving disagreements.

After Grand Coulee Dam was put into service, the author supervised the inspection of the spillway bucket of the dam where considerable damage was discovered.

On August 21, 1953 he was reassigned as Assistant Project Manager, thus again having an intimacy with the operation and maintenance of Grand Coulee Dam and Power Plant, which continued until his retirement in 1970.

Robert L. Telford, MWAK cofferdam engineer 1934-1937, now chairman of the board of Mason and Hanger-Silas Mason Co., Inc. reminiscing with the author on August 9, 1986.

"If nature could do it why not man??- William A "Billy" Clapp.

"You should see "Billy" Clapp and have him relate the story about a proposed dam at Grand Coulee to irrigate this basin".
   W. Gale Matthews to Rufus Woods.

"Baron Munchausen, thou wert a piker!" - Judge R. S. Stiner to Rufus Woods by letter ridiculing Woods story in the Wenatchee Daily World.

"We are only trying to get three blades of grass to grow where only two grew before". - Frank A. Banks at Congressional hearing.

Richard L. Neuberger, Portland, Oregon interviewing the "Three Starters," Gale Matthews, Rufus Woods and Wm. M. Clapp at lunch at the Green Hut Cafe in Coulee Dam. December 22, 1945.

James O'Sullivan being congratulated by Interior Secretary J.A. Krug in recognition of his long and untiring efforts in promoting the Columbia Basin Project and upon having O'Sullivan Dam named in his honor. It was formerly Potholes Dam. September 28, 1948.

CHAPTER I
GENESIS — IN THE BEGINNING

INTRODUCTION

Conceived in a time of hardship and humiliation and undertaken in adversity, Grand Coulee Dam would become the largest masonry structure ever built! When it was undertaken, the great pyramid of Cheops was still the largest structure on earth. The State of Washington was only one generation old and its homesteaders and pioneers in its midsection drooled for the bountiful waters of the great Columbia River which ran unfettered through its deep canyons to the sea—waters which, if brought onto the land, would make its semi-desert bloom like a rose. Most of the homesteaders had moved to less arid lands from whence they came and so the area south of the Columbia River in Central Washington was sparsely populated and the lands to the north of the river lay within the Colville Indian Reservation—also sparsely populated.

In this environment the visionaries of the Columbia Basin Project proposed damming the Columbia and pumping the water up into the Grand Coulee for irrigating the basin lands beyond.

Before moving there in April of 1950 I had not heard the real story of the struggle of the people of Ephrata in behalf of the Grand Coulee Dam but it was soon related to me by two of the citizens who saw it come about. Mrs. Ruby G. Wilson rented a room to me before I moved my family and household there, and I soon became acquainted with Mr. W. Gale Matthews who had come to Ephrata in 1909 where he established his Title Abstract business. The idea for damming the Columbia was first mentioned, according to Mr. Matthews, at a meeting of friends in the office of attorney William M. Clapp in late spring of 1917. Present also were A.A. Goldsmith of Soap Lake, Paul D. Donaldson, Mr. Matthews and possibly Judge Sam B. Hill. Agricultural development was the topic—but Donaldson related Dr. Landis's theory of a blockage of the Columbia below the head of the Grand Coulee that had diverted the river. Mr. Clapp then spoke up, suggesting that, "If nature could do it, why not man?" It was a local idea until Mr. Rufus Woods, publisher of the *Wenatchee Daily World* appeared in Ephrata in mid-July 1918 look-ing for a story. He interviewed Mr. Clapp—got the story of his lifetime—and published it in the July 18, 1918 issue. The idea got attention *and* support. Mrs. Wilson told me of the struggle of the townspeople contributing a few cents per week each to boost "The Dam Project" over the years of the twenties and thirties. They told, too, of the devotion and struggle of Jim O'Sullivan who spent his life's energy in the struggle for the project.

Their fervor was rewarded and their struggles and successes were recorded. Billy Clapp Lake and O'Sullivan Dam—both significant features of the Columbia Basin Project—were named in recognition of the contribution those two made in bringing the Grand Coulee Dam into reality. Those visionaries and their colleagues enlisted the interest of the state and federal officials in the scheme and preliminary investigations as to the feasibility of the idea were undertaken in the 1920s by the planners in the state and by the U.S. Bureau of Reclamation and the Corps of Engineers. Since irrigation was a paramount purpose of the proposed development, the assignment of primary interest in it rested with the Bureau of Reclamation at the national level which agency took the lead in cooperation with the state through its Columbia Basin Commission. The Bureau made an initial report on the project to the Commissioner of Reclamation on January 7, 1932. Controversy arose as to how best serve the lands—whether by a long, gravity canal from the lakes in Idaho, or by the plan favored by the local visionaries. But the pumping plan prevailed. By the early 1930s the nation was in the throes of severe economic distress, with industry at a near standstill and unemployment rising rapidly with its accompanying suffering and unrest.

Franklin D. Roosevelt was elected president in November 1932 and both local and statewide efforts were advanced seeking committments from him for the start of construction of the Grand Coulee Dam and the Columbia Basin Project. But demands for economic assistance and public works were nationwide and this was a huge vision. The State of Washington in strengthening its bargaining role for the project put forth $377,000 for continuing studies, surveys, investigations and specifications for a low dam. And on July 16, 1933, a symbolic driving of the first stake for the undertaking was had at the damsite— President Roosevelt having promised $63,000,000 to start the work as a P.W.A. (Public Works Administration) project.

Meanwhile, engineering investigations and studies and preliminary designs had been accelerating in the office of the Bureau's Chief Engineer in Denver, Colorado. The granite bedrock at the damsite was an excellent foundation for a large concrete dam though it was covered with a deep mantle of earth, clay and boulders. The river was second in size only to the Mississippi in the 48 states, and it was subject to large fluctuations in flow and level at the site, but it was a good, safe site for a high dam—how high was the question. The level of the backwater where the river crossed the Canadian border was one controlling limit—it was finally adopted as controlling. However, some advocated immediate negotiation with Canada such that the ultimate height of Grand Coulee Dam would raise the reservoir sufficiently to include the Arrow Lakes in British Columbia.

Electric service in Eastern Washington was provided by private utilities and hydro-electric generation plants with excess capacity, existed at Rock Island Dam on the Columbia, on the Spokane River and at the outlet of Lake Chelan. Also utilities had filings for additional dams on the Columbia and its tributaries. But at the time the need for large additions of electric power was not envisioned by many, if any. Nevertheless, the decision was made to construct Grand Coulee Dam to a height nearly 200 feet less than the planned ultimate, and to install power penstocks for 18 generating units but to build the structures in a manner to facilitate the raising of the dam to its ultimate height when that might be determined. Construction thus could proceed.

THE BUILDER

The builder of Grand Coulee Dam, the United States Bureau of Reclamation was an agency of the Department of the Interior, established by act of Congress in 1902. The ''Bureau,'' as it was called, was to operate in the portions of the United States west of the 100th meridian. Its purpose was to speed up the development of the West by construction of storage and irrigation works in the arid and semi-arid lands under suitable repayment arrangements with the benefitting water users. When I reported for work with the Bureau in 1931 it was a small agency, even by the standards of that day. It was located in Washington, D.C. and headed by a commissioner, Dr. Elwood Mead, who maintained a small administrative and liasion staff there. The planning

and design activities and the supervision of construction, operation and maintenance for the Bureau, however, was the delegated responsibility of its chief engineer—then Mr. R.F. Walter—and headquartered in Denver, Colorado. Mr. S.O. Harper was the assistant chief engineer, Mr. John L. Savage the chief designing engineer, with W.H. Nalder as his assistant. E.B. Debler was the planner and hydraulic engineer, as I recall, and L.R. Smith was the chief clerk and George Evans the chief of mails and files. A small legal detachment was also present. The total staff of the chief engineer when I started there was about 210. At that time the Bureau had already constructed 17 dams over 100 feet in height and its engineering capability was world-renowned. Boulder—later Hoover— Dam was under construction. Owyhee Dam was completed and soon the Bureau was one of the foremost engineering organizations in the world. Before the decade was out the Bureau would have built or was building the five largest concrete dams in the world. The undertaking at Grand Coulee would be in good hands!

GRAND COULEE DAMSITE AND VICINITY

In 1933 the Columbia River was a pristine river and its work of reducing to sea level the land that it drained had not been delayed by any works of man except for Rock Island Dam near Wenatchee which had just been constructed by a private utility. At the site for Grand Coulee Dam were but two habitations, in one of which Charles Osborn was housed and Sam Seaton and his family in the other, and two other families lived some distance away. Seaton's ferry, a cable-operated ferry of very limited capacity, was the means for crossing the Columbia. Traffic was sparse—travelers crossed enroute to Nespelem and the Tribal headquarters of the Colville Confederated Tribes, or drovers of sheep to or from summer pastures on reservation lands and few others.

Paul Bickford visited quite a bit with Charley Osborn as he came into the field office late each afternoon and one day they got to talking about his life out here on the riverbank all of those years alone. ''Did you ever think of marrying?'' Paul asked.

''Well,'' Charley replied, ''I did give it some thought once. You know, living out here next to the reservation I had some good friends among the Indians and once one of the sub-chiefs suggested that I take one of his daughters as my wife and mentioned the dowry of 27 horses. But nothing came of that,'' he said.

(cont. on p. 24)

Charles Osborn lived here undisturbed from the 1880s until he gave up his homestead for the construction of the town of Coulee Dam. April 1934.

Sam Seaton ferrying a customer across the Columbia River at the site of Grand Coulee Dam. Seaton's home is visible behind the clump of trees on the right—the east bank of the river. At the damsite the Columbia flows north. March 23, 1934.

During the War the parapets on the dam were of wood—those beautiful aluminum railings being deferred for aircraft. Here is a passing scene—Joe Hodgen, a wool grower from near Adrain, leads his flock of sheep across the now-completed dam crest enroute to the summer pastures on the Colville Indian Reservation. In years past they rode the Seaton ferry. Later they were to be hauled by truck. 1943.

Diamond drill crew took a thrilling ride when this mass of earth suddenly dropped beneath them. February 21, 1935.

Part of the first crew at the Bureau office in Almira. Jake Niemen, Don McLaren, Dick Stevens, Fred Berry, Robert Strubel, _____, Duke Ellis, Don Flanagan, Oliver J. DeSpain, Alex Blankevort, Earl Holbrook, Harold Wollenberg, Carl Berry, C.B. Funk, C.G. Rossman, R.W. "Dick" Smith, Edward L. Greene, Frank Wald, _____, Johnny Walker, R.T. Sinex, P.R. Nalder, John Breedlove, Jack Sowle, Walter Mackey, Mary

Cole (N.R.S.), Paul N. Bickford, Oscar W. Dike, Luther E. Cliffe, Harry Wilbert, Grant P. Gordon, Frank Phipps, C.M. "Baldy" Cole, Tom Heatfield, Pat O'Malley, Raymond C. Smith, Seth Norris, J.E. "Ernie" Hill, Althe J. Thomas, Ellsworth Mattson, Fred A. Saalbach, Walter Soule, Carl C. Scott, Ted R. Anderson, Barry Ware. March 30, 1934.

When Paul then querried, ''Why not?'' Charley kind of chuckled and said, ''You know there wasn't a good horse in the lot!''

It was a primitive setting; the stage for drastic change! And the roads to the site were mere unimproved wagon roads. The nearest improved road, State Highway 2 crossed the state, passing 20 to 30 miles from the site connecting the rural villages of Wilbur, Almira and Coulee City. But by August 4, 1934, the state built an improved road from Coulee City to the head of the Grand Coulee. The population of those towns totalled 1,496 in 1930 and the entire population of adjacent Grant, Douglas and Lincoln Counties was only 25,103 and the environment was strictly agricultural. Ranching and scattered dry-land cultivation of wheat and the servicing of such was the economic base. Electric and telephone service were available in the towns by then. The town water systems were scaled to the needs of the populations. Also a branch line of the Northern Pacific Railroad extended from Spokane to Coulee City, some 30 miles from the damsite. From Coulee City the level terrain up the bottom of the Upper Grand Coulee to its beginning at the Columbia River Canyon provided easy access but from that point at elevation 1500 the steep grade dropped down to the Columbia River at elevation 935 at low water in a matter of less than three miles. The ridge lying between Wilbur and Almira and the site rose above elevation 2500 and so while closer, forestalled the routing of heavy traffic from those points.

SITE SURVEY AND INVESTIGATIONS

With the funds advanced by the State of Washington and the president's commitment to proceed with the work, the Bureau of Reclamation assigned the task to Mr. Frank A. Banks, an experienced, very capable civil engineer who, as the construction engineer, had just finished the job of constructing Owyhee Dam in Eastern Oregon. It was at that time the highest dam in the world. He was assigned to the Grand Coulee Dam Project on August 1, 1933 and promptly gathered a nucleus staff (with headquarters in Almira until March 22, 1935.) Experienced personnel from the state, men who had worked on Cle Elum Dam and Rock Island Dam, from the State Highway Department—and from the engineering colleges of the state, bright,

young graduates were available for the asking. Among the latter Philip R. Nalder rose to the position of project manager for the Columbia Basin project—as did Dick Gray after Bill Rawlings retired in 1970.

Staffing, surveying and mapping of the site and investigation of the foundation and over-burden and fixing of the location of the dam were the high priority items and soon the established control surveys were extended outward, ultimately covering most of the Columbia River corridor from the Canadian Border to Oregon. Diamond drilling methods were extensively used for bedrock exploration, defining the depth of the overburden and the location and character of the granite bedrock. Trenches were blasted in the exposed granite along the selected axis of the dam above the over-burden contact to determine the depth of surfacial weathering. A search for the best sources for the concrete coarse aggregate and sand needed to build the structure was urgent. Grant P. Gordon, a civil engineer transferee from the Cle Elum Dam directed the contracted diamond drilling and the materials investigation crews who were directly employed by the Bureau of Reclamation. The surveys were under the supervision of C.M. Cole, recently employed on the Rock Island Dam.

Charles B. Smith was office engineer but resigned in March 1934 and returned to the Corps of Engineers. Arthur Dysart, an architect, was engaged for town and housing planning and design. Party chiefs to head survey parties included Carl Berry, Fred Berry, O.J. DeSpain, E.J. Niemen, Paul Piper, Harold Sheerer, Raymond C. Smith, A.J. Thomas, H. Wilbert and Ross K. Tiffany Jr. The chief clerk was C.B. Funk.

Mr. J.H. Miner, a civil engineer, arrived April 21, 1934 to the vacated office engineer position—he was well versed in Bureau of Reclamation policies and procedures, having worked on its projects in Colorado in earlier years. A.F. Darland, an electrical engineer, was employed on April 17, 1934 as field engineer—he had been on the consulting board of the Columbia Basin Commission and been employed by the city of Tacoma on its hydroelectric developments. F.J. Sharkey, the assistant office engineer arrived July 16, 1934. The surveys and data gathering for the dam, camp and road and railroad down the hill from Grand Coulee to the river and the control surveys began with this force. To the above-named and pictured were added rodmen, levelmen, instrumentmen, axemen, well diggers, draftsmen, tracers—many of whom were college graduates from the engineering

Grant P. Gordon, civil engineer—materials and foundation investigations chief. December 5, 1938.

C.B. Funk, chief clerk. July 1936.

E.D. "Dusty" Rhoades, diamond driller, foreman and superintendent—worked the dam foundations from start to finish; a keen observer. After a long meeting of supervisors he was asked if he had anything to say. His reply, "I was invited here to listen, and I've done a hell of a lot of it!" March 9, 1949.

colleges of the state. Some of those I recall who were added a bit later were Ed Greene (a very capable engineer), Ray Kerfoot, Layton Johnson, C.B. Cox, Robert McGowan, Harold Bock, Howard Q. Clark, Mr. Sharp, M.M. Smith, Dick Miller, Bill Gillogly, Al Landbeck, Walt Soule, Frank Maynard, Dan Hannah, George I. Owens, J. Sumner Smith, Gail M. Clevenger and Wendell Clark.

Initially the state of Washington furnished the vehicles and equipment and footed the bills. By April 1934 all of the payroll was transferred from the Columbia Basin Commission to the Bureau.

At the damsite headquarters—Almira—plans were quickly drawn and determinations made for the acquisition of land required for the construction of the dam and related features. A permanent camp for use of government staff would be built on the left bank, as viewed looking north, just downstream of the dam, while the contractor would be situated on the much larger benchland on the right bank where housing, shops and work areas would be needed. Suitable storage and marshalling areas were also envisioned and acquired in the floor of the Grand Coulee for use of the contractor. Negotiations with the state put the design responsibility for a permanent bridge across the Columbia, 3700 feet downstream from the dam with the State Highway Department.

Concurrently, in the Bureau's Denver office, designs and specifications for the "lowdam" and power plant matured, along with designs and specifications for the 30-mile railroad from Coulee City (Odair) to the left power plant, (N.P. Railway Co. engineers were engaged to locate the railroad), for the highway from the head of the Grand Coulee to the planned bridge across the river, and for the grading, for the water and sewer system, for the houses, office building, schools, dormatories and related features of the government camp. The government would procure and furnish the cement, machinery, gates, penstocks, reinforcing and structural steel, piping systems and other items required as part of the dam and power plant so careful scheduling of the anticipated needs was critical. But the so-called "critical path method" of scheduling had not yet been named. Designs and specifications were completed as timing for procurement dictated. The contractors for construction of the various features typically were to furnish equipment, supplies and labor needed to accomplish the job.

Close coordination between the job and the chief engineer's office was initiated and from that office came the details on drawings by the thousands. Availability of cement in the quantities that would be required and the transport of it to the dam was looked upon as a possible controlling factor in the contractor's rate of progress since cement mills were not numerous in the Northwest.

To provide employment and to hasten the start of work at the dam a contract was awarded to David H. Ryan of San Diego, California. He subcontracted with Goodfellow Bros. of Wenatchee, Washington, E.W. Ross of Los Angeles, California, McCutcheon Co. of Los Angeles and Rowland Construction of Seattle for removal of 2,000,000 cubic yards of the overburden from the damsite and actual excavation started on December 13, 1933, but notice to proceed was not received by Ryan until January 2, 1934. Addision Miller Co. built groups of tarpaper-covered shacks to house the workmen and operated a mess hall on the site. The "3 Engineers," a private firm, set up a 500-kilowatt diesel electric plant to supply electricity.

Ryan's contract included excavation on both sides of the river with disposal upstream from the dam where it was hauled by trucks. The four shovels were gas and diesel powered, having combined capacity of 6¼ cubic yards and the 14 dump trucks totalled 64 cubic yard capacity, typical of construction machinery of that time. As excavation proceeded the underlying clay proved to be indurated and stood vertically faced, the result of it having been compressed by the overriding continental ice sheet after it had been deposited in the glacially-fed lake in the distant past. One sub-contractor sought reclassification of the material as rock, but resumed work as bid when he found that his contract did not provide for rock excavation. In the early work conditions portending troubles in the future developed as excavation removed the toe support for higher lying deposits of the overburden and severe slumping occurred spilling the disturbed earth down into the excavated area. And on one occasion the crew of diamond drillers working thereon got a scarey ride on the moving mass. This massive slide required relocation of the railroad and highway and involved much additional excavation.

Bids were sought for other features as rapidly as designs and specifications were completed in order for employment of as many as could be used of the anxiously waiting work seekers. Soon contractors were on the job building the railroad (David H. Ryan), the road down from the Coulee floor to the riverbank (Crick & Kuney), grading the townsite and its streets (S.H. Newell), constructing the water and sewer systems (Arcorace), houses, the garage and fire station, dormatories, office building, school, post office, the piers for the highway bridge

(Western Construction Co.), and a large warehouse for storage of government supplied materials and equipment. The dirt was flying!

THE "LOW DAM" — ENGINEERING FEATURES

Field investigations and drillings had disclosed that the bedrock of the damsite had been eroded in glacial or preglacial times down to about elevation 880 feet above sea level across the river floor for a distance of nearly 3000 feet then rising gradually though somewhat steeper on the left bank such that with a water level set at that at the Canadian border the dam would span nearly 4300 feet in length. Surface geology and terrain showed certain weathered zones (slight ravines) that later excavation of the overburden proved were continuous reaches of less than perfectly sound granite. But for the most part the cores removed showed a very good foundation in prospect. Design studies advanced, and specifications were prepared for the dam—it would be built to elevation 1116; it would accommodate the future cofferdamming of the river if and when the dam was raised to its visualized ultimate height; its foundation would be consolidated by grouting; drainage holes would be provided for reduction of uplift pressure in the foundation; its spillway would pass the anticipated flood flows of the river; it would have a grouting and drainage gallery about ten feet or more above the foundation; and inspection galleries at 50-foot intervals in elevation; outlet gates and tubes to permit discharge of storage water were included; a powerplant on the left bank containing nine units with penstocks 18 feet in diameter and the penstocks and the foundation for a similar power plant with nine units on the right bank of the river. Additionally with the possibility that the dam might be raised to its ultimate height before the passage of many years bid items were included for the transfer of title to the government from the contractor for its aggregate plants and cement handling facilities, conveyor systems, rolling stock, camps, power, water and sewer systems all of which would be useful in the event the structure were to be raised. In all, though, it was but "half a loaf!" Bids were opened June 18, 1934 in Spokane, Washington—a gala occasion for the "Dam Boosters!" The low bid was made by a joint venture, the principals of which were the Silas Mason Co. of Louisville, Kentucky and New York, the Walsh Construction Co. of Davenport, Iowa, and the Atkinson-Kier Company of San Francisco and San Diego,

Contractor David H. Ryan, equipment and crew for west abutment initial excavation contract. April 18, 1934.

California. The low bid was $29,339,301.50, about 15% below the Six Companies' bid. The event was celebrated of course! This was a big job for its time, though Boulder/Hoover Dam had gotten underway a few years earlier. But before the contract was awarded, Mr. Banks had met in July with state officials in Olympia about the high dam possibility.

Finalizing contract matters, including performance bond to assure completion of the work had to run its course, and the contract was signed on August 18, 1934 and notice to proceed issued on September 25, 1934. But the contractor, MWAK[1], it was called, had not been idle. The partners were all experienced in heavy construction work and they had varied types of needed skills. Too, they counted the days that remained before the Columbia River would be rising to flood stage the following spring—time could not be wasted. Promptly, designs for the first cofferdams were made and refined as more data arrived and the contractor's camp (to be known as Mason City) was started. The office, the homes, the bunk houses (for single men or those whose families remained at home), the mess hall, the hospital, the hotel, the restaurant, the store and the recreation hall together with the utilities to serve the "spread" of buildings, took rapid shape and occupancy.

The camp water supply was drawn from the Columbia about a mile upstream from the dam. Houses were prefabricated in panels and trucked out to the site from Spokane—no chimneys were provided since electric heat only was used, and by November 1934 some 254 buildings were under construction. With notice to proceed, the pace accelerated and excavation of the overburden moved center stage. By October 28, 1934 the contractor had built a temporary pile bridge across the river and was extending the construction railroad skirting the government camp, blasting out a tunnel and suitable grade down to the river at the north end of the campsite. There a bridge was built across the river, intended for rail service to the right bank shops and work areas, but subsidence of part of the bank and shop area prevented it. The bridge did serve for truck traffic crossing the river, but never a train.

Electric power to serve the area's growing needs was on the way from the Washington Water Power Co. system at Coulee City with connections to its Spokane River and Lake Chelan hydroelectric plants. MWAK was building a 110,000-volt line to the site. It was energized on November 18, 1934.

---

1 But the Thompson-Starrett Co. had also joined the MWAK after the bid.

Principal partners of the MWAK Co.—Silas Mason, Guy F. Atkinson, Tom Walsh Sr. and Elmer L. Kier. 1935. (Courtesy of the Guy F. Atkinson Co.)

While that ink was still wet—so to speak—and that notice to proceed on the Low Dam not yet delivered, the basin boosters for irrigation of the Columbia Basin Lands became concerned with the future of their dreams. With a Low Dam "generating power to be unsold" so the opponents claimed, it may be many years before their dream would be realized. So with renewed vigor, new recruits came to their cause because of the still present high rate of unemployment in the nation and the fact that the construction of the dam provided work in manufacturing; in the foundries of the South, in the steel mills of the Ohio River area as well as in the West and the public was getting its money's worth *Reclamation Expenditures Were to Be Reimbursed!* President Roosevelt visited the damsite on August 4, 1934 and he was impressed—there was an enthusiastic crowd "way out from nowhere" and he began to really appreciate the point of it all. The High Dam Boosters got his ear! The "word" seemingly was passed and the chief engineer of the Bureau of Reclamation began studies on the "how" of changing the contract with MWAK to accommodate such a major change as the boosters wish would require. Things moved fast in those days when urgent means were taken to get jobs for the many. The chief engineer's designers thought the problems they foresaw should be reviewed and the chief engineer, Mr. R.F. Walter, met with the Commissioner of Reclamation, Dr. Elwood Mead in December 1934 to review the changed economic conditions and engineering and construction features in the two-stage plan on which MWAK had started. While the Low Dam was an excellent work relief measure, it was not the best choice of site for a low-head power plant because of depth to rock and length of dam for the head developed and slides yet in prospect substantially increased excavation requirements and the cost of the auxiliary features (railroad, highway, bridge, camp, cofferdams and excavation) were about the same for the Low Dam as for the High Dam. Also there were unresolved questions about the engineering problems of the construction joint between the two stages. Too, the planned installation of generators to operate at the low head and the high head envisioned could be costly. Lastly, the drought in the midwest had caused the migration of many farmers from marginal lands and they sought new opportunities. The Commissioner so reported to Secretary of the Interior, Harold L. Ickes. But on the job,

the work went on—seemingly on its plotted course. But shortly, we knew that a major change order was being negotiated. When concluded with MWAK Company, (the order was dated June 5, 1935), it, with adjustment of compensation document, provided that the work under the contract would be changed from a Low Dam to the base for the High Dam. Principal changes were: increased excavation, change in shape and dimensions for the dam, construction of longitudinal contraction joints, shortened spillway, eliminated permanent cofferdams, construct openings for all penstocks to elevation 990, add cooling and grouting systems, eliminated powerhouse construction above turbine floor, include excavation for pumping plant and install paradox and ring-follower gates for the outlet works. It increased the contract cost by $7,000,000. By act of Congress, August 5, 1935, the project was identified as Grand Coulee Dam, but included irrigation, navigation, flood control, reclamation of land and generation of electricity to assist financially. Also, it ratified all contracts and change orders previously made on the project.

Three "Big Chiefs" of the "Reclamation Tribe" in headdresses, courtesy of the Nez Perce Tribe at the festivities of the bid opening for construction of Grand Coulee Dam. Frank A. Banks, construction engineer, R.F. Walter, chief engineer, Dr. Elwood Mead, Commissioner of Reclamation. June 18, 1934.

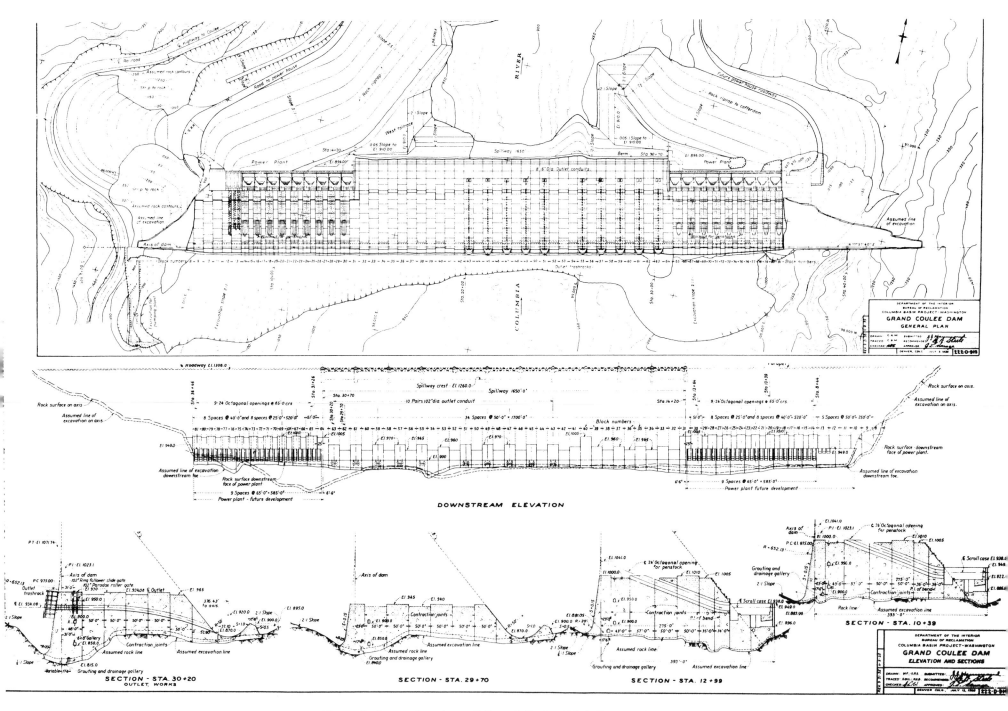

General plan and sections of Grand Coulee High Dam. July 1935.

The high dam at Grand Coulee as visualized in preliminary designs and analyses would be a straight gravity dam with roadway at elevation 1311 feet above sea level with the downstream face sloping 0.8 horizontally to 1 vertically and with the upstream face vertical above about elevation 1000, below which the face would slope slightly upstream. As built, the dam was 4300 feet long, 553 feet high and over 450 feet wide at maximum section. The spillway for the passage of flood flows would have a capacity of 1,000,000 cubic feet per second and the crest of the gates thereon would be at elevation 1260 providing 30 feet of surcharge and 21 feet of free board against overtopping. The energy in the water passing over the spillway would be dissipated in a plunge pool or "bucket" 70 feet or more in depth, at the bottom of which the downstream face of the dam terminated in a cylindrical section of 50-foot radius. The spillway bucket so formed, with invert at elevation 870, would redirect the plunging water back toward the surface of the river, thus further reducing the unspent energy.

Because of the great mass of the structure and the vast store of heat in the structure from the hydration of the Portland cement in the concrete, it would be desirable to remove the stored heat by cooling the mass down to about the prospective ultimate temperature of the structure. Without cooling, it was projected that 200 years may elapse before the structure stabilized. The concrete mix would be designed to obtain maximum strength for the cement used and the mixing should produce uniformity of the product. The concrete would be placed in formed blocks of such size that between the successive layers in any 5-foot lift, good mixing between successive layers in any pour would be assured. Typically the concrete then would be placed in formed columns 50 feet square with the formed surfaces keyed to facilitate bond and the transfer of stress between the columns. The joints between the columns or blocks would be filled with cement grout after the mass had been cooled. The treatment of the foundation and provision for galleries were as described in the discussion for the Low Dam. Sixty discharge outlet tubes would be provided, gated in tandem, two tubes to a block and located at 100-foot vertical intervals.

The power plants and the spacing of units therein govern the location of the contraction joints (the formed surfaces) in the portion of the dam adjacent the power plant. The spacing of units being 65 feet, the blocks in the dam were 40-feet wide for the blocks containing the penstocks and 25 feet for those adjacent. Contraction joints paralleling the dam axis were at 50-foot intervals but were staggered so that continuous surfaces would not extend through the structure in that direction. The planning for power plant capacity in the late 1920s and 1930s seems to have been based upon very modest growth in demand with justifications based on the prime power available with little consideration for the value of peaking power. The units were sized with a rated capacity of 108,000 kw and the turbines were rated at 150,000 h.p. When the Low Dam specifications were issued, space for 18 units was required. These turbines and generators which were being contemplated in the early thirties were nearly three times as large as any then in service. Those for Boulder/Hoover plants for example, were rated about 83,000 kw and the turbines rated at 115,000 h.p. but they had not yet come into service. I mention this as a reminder of the progress then being made in the entire electric utility enterprise in this country.

The plan for the pumping plant at the dam envisioned 12 pumps with a capacity of 1600 cfs each. The lift from the reservoir at elevation 1290 to the full balancing reservoir (Banks Lake) at elevation 1570 would be 280 feet, but with the reservoir at maximum drawdown at elevation 1208, the lift would be 362 feet. The plan contemplated using not more than 10 pumps at any one time—the other two as spares in the event of failure during periods of peak demand at the ultimate development of the million acres of the Columbia Basin Project. When planned in the thirties, the possibility of Canadian storage in the headwaters of the Columbia was not something that could be counted upon. Then it was anticipated that the pumps would be operated primarily during the seasonal floods using secondary power.

The pumping plant would rest on the downstream slope of a wing dam, an adjunct of the main dam. The wing dam to be constructed as a part of the High Dam contract would be 588 feet long with a height of 150 feet (plus the foundation.) The back slope of the wing dam would be 0.67 to 1.

BOARD OF CONSULTING ENGINEERS

Periodically the Bureau convened a board of consultants to review the progress of the work and to give valuable counsel and recommendations on specific technical problems encountered or foreseen in the construction of the project. The board consisted of Dr. Charles P. Berkey, Dr. W.F. Durand, Mr. D.C. Henny, Mr. Joseph Jacobs and Mr. C.H. Paul. Mr. J.L. Savage, the Bureau's chief designing engineer, usually accompanied the board to review the technical problems involved. Chief engineer R.F. Walter accompanied the board in June 1934, as did Mr. L.N. McClellan, Bureau chief electrical engineer at one session. Mr. Banks and other engineers, inspectors and geologists at the dam also met with the board to describe aspects of matters on the agenda and to answer questions and concerns of the board members. My assignments involved subjects with which the board had special interest, so I met with them on several occasions to report and to respond to their queries. I recall their keen interest in the cofferdams, instability of the overburden, results of observations of drainage from the overburden, foundation exploration by calyx drilling, excavation of the bedrock, grouting of the foundation, and the condition of the spillway bucket.

Dr. Berkey was the eminent geologist from Columbia University in New York City. He had a sharp, steady eye and it twinkled—he enjoyed his work and his associates and he was an effective teacher. When discussing the occurrence of some of the seeming incongruities of the terrain with huge boulders of granite and of basalt strewn about, he said without hesitation that ice had accomplished that task. When then asked how he knew, he calmly, but with that twinkle in his eye, answered, ''Well, if I were ice I would do it!'' Dr. Berkey had been the geologist with the Roy Chapman Andrews' exploration into the Gobi Desert of western China in the early twenties and his store of knowledge and his willingness to share his experiences with young and old alike endeared him to us all. And he didn't look upon his visits to the dam as being very isolated at all! Solace to the ladies in camp. Dr. Berkey maintained a continuing interest and contact of sorts with the geology of the dam and surroundings through one of his proteges from Columbia University. Mr. William H. Irwin came out to the dam during the summer of 1934 (?) and later returned after graduation as an employed geologist on the staff of the Bureau.

Dr. Durand too was an academician, the eminent head of the Department of Mechanical Engineering at Stanford University at Palo Alto, California. He, too, was broadly versed in the engineering field and his questions were clear, concise and on target. It was up to us on the dam to get the answers he sought there. He was the one who wrote the board's report in draft and in final. Mr. D.C. Henny was a successful consulting engineer from Portland, Oregon and had considerable experience in the design and construction of dams and other water resource projects. He had been an engineer with the U.S. Reclamation Service shortly after it was established in 1902. In 1904 he reported in the fourth annual report of that agency regarding the feasibility of irrigation of the lands in the Columbia Basin by long canal systems from the lakes in northern Idaho—he found the plan wanting! He was quiet and deliberative, was knowledgeable of the forces the dam would have to endure and he gave it his full attention. His son Arnold Henny and I worked together in the Chief Engineer's Office and played golf together, so I was fortunate to know ''Mr. D.C.'' before coming out to the dam. Henny was Holland Dutch. He died on July 14, 1935. Mr. Joseph Jacobs was a successful consulting engineer from Seattle. He was a quiet ''thinker'' and some of his major concerns were the numerous slides, the need for a tight foundation under the dams for the project and the control of seepage. Mr. C.H. ''Boss'' Paul of Dayton, Ohio was the board chairman. He had been employed by the Bureau on the design and construction of some of the then-large dams on the Snake River in Idaho in earlier years and so had known Mr. Banks well. He was employed on some of the civil works on the rivers of Ohio in the thirties, but I don't know the specifics. I recall that he had been out of the meeting when I was reporting to the rest of the board on the results of the foundation grouting, the large quantities of grout that were being injected into some of the holes and the fact that the cores from the diamond drill holes then showed clearly-defined and well-bonded multiple layers of grout, and that we were, by the grouting effort, actually lifting the dam and consolidating an increased area of the foundation. When he came in, he asked for me to bring him up to date—in brief—and when I mentioned the multiple layers of grout in the cores he said, ''Well, you are lifting the dam!''

Mr. Durand responded, ''Yes, that is what Mr. Downs was telling us—and we think that is to the good!''

I profited greatly from those opportunities to meet with the Board of Consulting Engineers and I never hesitated in responding to any question that they, Mr. Savage or Mr. Banks asked me.

It would be a gross error on my part to omit the role of the chief engineer and his officers and aides that designed and maintained technical control of the construction of Grand Coulee Dam and appurtenances. That office in Denver then was the foremost engineering office in the world in the field of heavy construction of water resource projects. And when Grand Coulee Dam was undertaken, that "office" was still building Boulder/Hoover Dam, but Owyhee and many others going back to its beginning, had their "engineering" from the Chief Engineer's Office.

I knew many of the men who gave the office its great reputation and I mention them only by name except for the chief designing engineer, J.L. "Jack" Savage, who was both the chief of design (and construction)—a separate office of construction supervision had not yet materialized.

R.F Walter was chief engineer and his first line support included besides Mr. Savage, S.O. Harper, W.H. Nalder, B.W. Steele, F.F. Smith, L.N. McClellan, E.B. Debler, Mort Day, H.R. McBurney, Sam Judd, Al Kinzie, Ivan E. Houk and R.S. Lieurance (to those two I was assigned), John Hammond, McConnaghey, Walt Blomgren, Bob Blanks and Arthur Ruttgers. Among us subordinates were: Dick Whinnerah, Dick Larsen, Arnold Henny, Ray Dexter, Merle McCleery, F.D. Kern, Fred Houk, Ed Green, J. Sorens, Scott Bair, Wally Waldoff, G. Manley, Bob Sailer (an outstanding, innovative structural designer), Kenneth Keener, Jim Ball, Bill Woolf, Charley Thomas, Walt Price, Lee Snyder, W. Holtz, Max Kight, Jake Warnock, I.A. Winter, "Bud" Wilbor, Lewis Workman, Clarence White, Carberry, Randy Riter, Francis Thomas, Ned Trenam, Emil Lindseth, John Parmakian, Les Bartch, Skip Noonan, Ickkie Silverman, Pete Bier, Jerry Ross, Church, Phil Noble, Ernie Schultz, Richardson, Oscar Rice, Bentson, H.W. Benton, Fred Wilhelm, Ralph Burkhart, Charles Burky, Harold Robbins, Charlie Masten (those beautiful drawings), Bill Beaty, Plumb, Gustafson, Harvey Olander, Chuck Rippon, Lovell, Hunt, Allan Davis, Ed Rose and many more whose names do not come back to me.

They saw problems, visualized a solution, refined it, designed it, put it onto drawings, and into the specifications with complete instructions on what it was to be and how it would work—and for the most part, it did! And they did all this well and on short notice. Their designs were not off the shelf items—Owyhee, Boulder and Grand Coulee Dams were each of the "one of a kind varieties." They kept in touch with us in the field and we with them.

"Jack" Savage, as he was known, gained world-wide renown for his design and consulting work with the Bureau and the governments

Board of Consulting Engineers at the dam in December 1934. D.C. Henny, Dr. Charles P. Berkey, J.L. Savage, C.H. Paul, Chairman, Joseph Jacobs, Dr. W.F. Durand, Frank A. Banks. December 1934.

Future site of Mason City, the contractor's camp on the far bank of the Columbia River. Contractor's workmen on early work at the dam were housed in the small camps on the riverbanks. Addison-Miller camp on near side, center of photo. The "Three Engineers" power plant in right foreground—see smoke. July 7, 1934.

around the world. Although he received many honors, he was a quiet, reserved gentleman, revered by many! Among his many honors was a "Special Award" from the Beavers at its 1961 honors meeting. Dr. Savage was a very kind and generous person and though he had no children he certainly was interested in them, and kids in many parts of the world knew him as "Uncle Jack." He was interested in their success in life and in their education. In that connection, he was quite generous in giving financial assistance and encouragement to numerous nieces and nephews and others for their college careers. He enjoyed good company and a good joke. Fred Sharkey and Dr. Savage drove over to Hungry Horse Dam from the dam once and enroute they got to telling jokes and laughed so hard that Fred almost wrecked the car—so Fred told me.

Professionally Dr. Savage received many honors, including the honorary degree of Doctor of Science from the University of Wisconsin in 1934, the John Fitz Medal, honorary member of the American Society of Civil Engineers and many others. But he was modest to an extreme. I recall his concern about publicity directed to his achievements when he had been serving on a board of consultants for the International Boundary Commission for the U.S. and Mexico. The press had barely mentioned Mr. Lawson (?) the U.S. representative and Dr. Savage thought that quite improper since he considered himself just one of the group of advisors. He had noticed a cartoon in a newspaper enroute from that assignment to Coulee Dam and he had clipped it to send to Mr. Lawson with an apology for "stealing" all the publicity. His note to Mr. Lawson mentioned his embarrassment and his discomfort like the ostrich in that cartoon that had swallowed a great Z-shaped bar of steel that had lodged in its neck. That was the first item for Dr. Savage's attention when he arrived at our home! With a person of such character it is no wonder that he had such a loyal, capable design organization.

CHAPTER II
RELATIONSHIPS

APPROPRIATIONS

Funds for the construction of the dam were appropriated by the Congress of the United States. In the days of Commissioners of Reclamation, Dr. Mead, Mr. John Page and Mr. Harry Bashore, the testimony seeking appropriations was brief and direct. These men who had come up through the Bureau and the construction engineers from the big projects of the Bureau were genuinely knowledgeable about the jobs they had to do, how they were planning to do it, and how much it was expected to cost. They were reputable professionals and their integrity was unimpeachable. So their appearances before the congressional committees were often very brief after their opening statements. I remember reading the testimony for the appropriations for the dam in one of those earlier years when the request was for something on the order of $10,000,000. Mr. Banks was responding to a query, "Is anything unusual about what you are undertaking out there at the dam?"

He replied that, yes, there was something unusual for the government to be doing. "Our camp for the employees of the government is located on land acquired for the building of the dam and there are open lands and very few residents in the vicinity, so we have planned to build a schoolhouse so that the children can be schooled—the building will cost $25,000." That is the substance of the discussion, not an exact quote, of course, but as I recall, that concluded that hearing and the money was appropriated.

When the agencies began to "go formal" with masses of data in support of the many programs and proposals, the Congress seemed to sense an "invasion of turf," and likewise began to expand the committee structures and the accompanying staff, so in later years the committee hearings went on for days and reading the record gave me the impression that "Those guys don't have any confidence in each other," which just might have been the case! But then too, it appeared to me that the Bureau had become politicized to an unpleasant degree!

LEGAL

When the project first got underway, legal matters for the project centered in Denver and a regional counsel was located in Portland—Mr. Ben Stoutmeyer. But in 1935, an attorney, Mr. Fred Hamley (later of the U.S. 9th Circuit Court of Appeals) was stationed at Coulee Dam. Paul Lemargie, Felix Rea and Bob Milne, who replaced Rea, later served at that station, so legal advice on the many contract matters for the project got good, prompt attention. And nobody went to jail! Actually, I do not recall many serious infractions of the laws, rules and regulations except for a case of acute "baseballitis" said to have happened before my arrival in 1934 when one of the fans on a "Sunday drive," hit a horse with the project station wagon he was driving while going to a ball game—well outside the scope of the project. One other instance I recall involved pilfering government property for private use.

I believe the most serious incidents involved the assessment of costs for the extra burden placed upon the Bureau because of fraudulent X-ray practices by the penstock fabricator.

Neither do I recall any investigation of alleged improper conduct in our relationships with the many contractors involved over all those years. I remember a discussion in December 1934 with Bill Allison who was assisting me when we were reviewing the contractor's time and attendance reports and billings for materials and supplies for constructing a tunnel under the big slide on the left abutment. The work was being performed as "cost plus" extra work. Bill commented, "You know, if there will ever be an investigation of this project, the first place they would look, would be the Extra Work." Recently, in 1985, the media pressed the apparent shortcomings in procurement practices in the Defense Department.

TOWNS

Towns grew like mushrooms beginning with the start of excavation for the dam in 1933. The Bureau had promptly condemned the properties for the damsite if negotiations with the owners were not successful. And there was very little land upon which to build except at the mouth of the Grand Coulee. The government camp—Coulee Dam—and the contractors' camp—Mason City—were not to be large enough to house the workers and the service establishments. Developers staked out lots on various ownerships and soon Grand

Coulee (Charles Howell, a hardware merchant, was the first mayor), Grand Coulee Center, Grand Coulee Heights, Grand Coulee Dam City (I think it was), Delano, Electric City and Osborne were identities on the previously bare landscape—grass, sagebrush and rock. Farther down the coulee, Basin City became a service station at a corner on the highway with a few dwellings nearby—good water from a spring was there. On the other side of the river, Lone Pine, Elmer City, Seaton's Grove and farther down the river, Bellvedere also took shape. Stores, a hotel, restaurants, theater, post offices, service stations, garages, pool and card halls, dance halls and rooming houses as well as houses and shacks rose up like tents seemingly overnight—and there were tents too. This was before trailers and mobile homes. Water was scarce and sanitation not modern and fires were frequent. And the dust was "knee deep," but when the fall rains came the mud was "knee deep." Somehow the workers and the businesses found quarters and by the end of 1935 there were eight towns and perhaps 12,000 people living within 12 miles of the dam.

Many large construction jobs leave ghost towns when the work is done, but at the dam the work is still not done and many people finished their working life there and retired there. So now, although very few of the shacks of old are still on the scene as reminders, there are many very nice homes on the shores and uplands of the river and the lakes. The population of the towns within 12 miles of the dam in 1980 was 16,000.

When the irrigation phase of the project got underway, it was necessary to vacate many of the town properties for the feeder canal and the reservoir (Banks Lake). But in 1934 no authority was "on the books" to acquire the undeveloped land. But Mr. Banks had thought of it! After the War with the planned influx of Bureau personnel to finish the "works"—by government force—an immediate shortage of housing in the area was foreseen. So the town of Mason City was greatly enlarged. The Brett Creek was diverted into a large concrete culvert and waste material from the excavation of the drydock was deposited there. The area including the old airport site was graded and made ready for the housing quickly demountable and transportable from the Vancouver shipyards. It has been said that there is "nothing as permanent as temporary" and so it is—some of those thin-walled houses are still in use, but most have been replaced or substantially rebuilt. Now the entire community of Coulee Dam as the now-incorporated area at the foot of the dam is named, has grown up with

Secretary of the Interior Harold L. Ickes and U.S.B.R. Commissioner John C. Page visit the dam. Mr. A.F. Darland, Mr. J.H. Miner, Mr. Page, Mr. Ickes and Mr. F.A. Banks. October 25, 1938.

many beautifully designed homes and with manicured surroundings greeting the visitor's eye. And great trees bless the town where once only sagebrush thrived.

In the government camp the Bureau forces maintained the properties and operated and maintained the utilities. There were painters, Les Whipple and Burt Whiting, and Willie Broden tried to keep electric service current, and there was Bill Butler the plumber, who told Ethel Hawkins one day, ''Well, you engineers' families may live in the nice houses, but we craftsmen get the most money!''

When the work on the dam began to wind down the Bureau saw the need to take over the contractor's camp—houses and commercial buildings. The Bureau was then faced with the disposition of the commercial properties. For the convenience of the residents, a grocery store was needed, and since no commercial venture looked promising for private capital, a cooperative was formed. The Coulee Dam Co-op was formed in July 1942 and provided basic food products for the community until it was liquidated in 1977. V.J. Peterson was very active in the organization of the co-op and served as president of the board for several years. Menno Alberts was the manger for a number of years and made the store a successful venture.

After the War the town grew and running the city was an added burden, so Mr. Banks employed a ''city manager'' from the East—Bob Garen. But the government town was fading as a Bureau institution because the work of building the features at the dam was coming to a close. In late 1947, I believe it was, Mr. Banks designated Fred Sharkey, C.E. Benjamin and me as a committee to serve in the role of a town planning committee. Our immediate function was to gather information helpful in the administration of the commercial properties of the town and to prepare data that might be useful in the transition of the town to incorporated status. In our efforts we visited a number of towns about the state and had the cooperation of Mr. Josh Vogel, director of League of Washington Cities. But our product was only an interim measure. Later Mr. George Shipman was employed to guide the governmental process to legislative authorization for the conversion of the town. And finally Gus Empie was employed to ''work himself out of a job'' by selling the properties.

The government sold the houses and lots in Coulee Dam and Mason City and is not in the housing business, most of the population are homeowners. When it seemed likely that the Third Power Plant

The camps. Mason City on the east bank and Coulee Dam on the west bank. Contractor's family housing in left foreground and beyond (south) of the K-shaped building (hospital). Single men's bunkhouses lay between the contractor's "store" buildings and the mess hall. The women's dormitory at left of Mead Circle at the intersection of Roosevelt and Ickes. The contractor's bridge in the foreground never carried rail traffic to the shops because of slides along the riverbank. June 1935.

would be authorized, I recall telling our "land master," Tork Torkelson to take the remaining unsold lots in the town off the market—to the consternation of would-be land speculators. Many of those lots were later needed for the new work.

CAMP LIFE

Well, what was camp life like in the thirties? First, there was the small camp to house the workers on the first excavation contract, the operators of the three engineers' small power plant, plus the cook and waiters of the mess hall and for the few Bureau employees at the site. The buildings were temporary—shiplap and tarpaper covering with six of us to each bunkhouse.

I shared bunkhouse #7 with Willie McKay, Don Schlapkohl, Ted Mann, and Hertice Marsh. Hertice was a dreamy architect. He was the one who snapped the padlock on the door while I was sleeping after working the graveyard shift. How could I forget him?

Some evenings the spirit of the camp picked up when "Red" Bird got too much into the spirit of the dime novel paperbacked western he was reading and grabbed his Colt from its holster and hurriedly put two or three quick slugs through the floor and shouted, "I got him." And that bunch in bunkhouse 13 also did target pracice with their .22 rifles or pistols with the targets drawn on the rear wall, which faced the river bank. The latrine was located farther downstream, beyond the last residential cabin. The path ran behind the cabins—my first experience with the "Critical Path Method."

Then came the "Government Town" with its small houses and raw yards with the dirt in the air, the unmuffled trucks polluting the air, the dynamiters (Blackie Armstrong) giving you notice of intended shots—shouting, "We're gonna blast!" BANG!!! And the many steam-powered rigs spouting steam, blowing whistles for the change of shift, the dozens of pile drivers pounding on steel, and the workmen shouting! My mother-in-law sent us a clipping from the Nampa, Idaho *Freepress* of a picture of the great concentration of crane booms, pile drivers, steam rising and spouting—and across it she had written, "Is it *really* like this?!!" Yes, it was!

But—oh environmentalists—where wert thou? And then for a few hours on the sabbath it came to a halt—for repair to the rigs and servicing the equipment—and it was *quiet*! If you happened to be asleep that shift, you woke up with a start, wondering, "What was that?!!" The quiet was deafening.

The ladies had two problems with which to cope; rumors and too few doctors. At a gathering at the camp one day the resident wives were asked by the wives who had lived in the cities, "How can you live here? How can you be happy and get along?"

"Keep your bowels open and your mouth shut," was the sage advice.

In those first days, getting groceries was a long walk to the store, but Dube Brothers delivered your order. Or you could bum a ride with a friend who had a car. Since I was still paying off college debts when I got married, we didn't get our first car until 1939, a '38 Chevy with knee-actioned front end.

Workmen from the bunkhouses over in Mason City sought recreation in pool and pinochle or poker at the recreation center and then came the pinball machine—I am surprised as I write this that I cannot recall one instance of anyone throwing a hip or vertebrae out of place from the distortions I witnessed being exercised in body English, trying to influence the machine without tilting it; the course of the pinball as it careened down the slope missing the holes that counted. Injuries of that type seemingly were all "job related." It was a spectator sport, too, the forerunner of the video games in the arcades of the 1980s. Those seeking greater excitement were known to have been seen in the bars, card rooms and dance halls and there were less advertised upstairs rooms on "B" street in the town of Grand Coulee at the top of the hill, off the "Government Reservation." The sheriff of Grant County kept deputies near at hand as a matter of economy—it was just too far from Ephrata the county seat to be called, considering the frequency for "custodial care" among the celebrants. Gordon Nicks was a deputy sheriff beginning in 1934, then sheriff in 1938. I never had any official contacts with Gordon until years later when he was a county commissioner. In Mason City and in Coulee Dam and on the highways, the calm and order was observed by the Washington State Patrol. Floyd Hansen was the first resident deputy sheriff and Paul Lemargie and Charlie Zack were justices of the peace.

But some preferred the city life on weekends and headed for the big city—Seattle—some even riding with "One-Way Lewis." He had few customers indeed, who ever made a round trip with him at the wheel. They were willing to risk their lives on the dam, but they abhorred the idea of dying of fright. Bill Allison once rode to the top of the hill with Lewis who was driving on the extreme outside left of the road against oncoming traffic not mindful of the survey stakes, which caused a blowout. But he kept at the wheel till he got up to the garage in Grand Coulee to have it fixed. There Bill said, "Mr. Lewis, this is as far as I go!"

One Sunday morning just before Christmas 1934 Don Schlapkohl and I went to the little post office at the dam to get our mail and since it had had a late arrival, it was not yet sorted. Harry Robbins, the postmaster, invited us to come back into the "behind the scenes" portion of the office where he had the warmth of a stove to share with us. He continued to toss the mail sacks onto the counter for sorting as we talked. When he plunged both arms into one sack he got the surprise of his life. In fact, he withdrew them so suddenly he nearly fell over backwards! When he carefully examined the contents we saw that someone had mailed a Christmas turkey—carefully packed in a carton, but had left the feet protruding about eight inches! Harry had grabbed those cold, sharp-nailed feet!

One cold, wintry morning, Don Flanagan came home from a night on graveyard shift and stripped off his work clothes. Barefoot and clad only in his long johns, he stepped out onto the porch to get the morning paper. His wife Calla had gone to work, and so he was locked out of his cabin when the latch clicked behind him. His kindly neighbor saw his plight and called to him to come on over and get warm while Calla could be contacted for the key. Living in the courts, their neighbors were close and numerous. Later, while recounting the incident, the neighbor remarked that no thought had been given as to how it might have looked to others in the neighborhood when Mr. Flanagan was invited over, but the thought soon hit *her*, "How am I going to get him home?"

Of course the camp had its problems with kids and dogs, and Bert Hall spent many an hour in Mr. Darland's office on behalf of his staff, responding to complaints leveled against either their beasts or offspring.

During the War many residents had large Victory gardens where the bunkhouses had been.

Dr. K.L. Rao, who had been teaching in an engineering college in England, visited the dam in 1941 and from him we learned of the privation that nation was suffering, as the War had been raging there for two years. He craved fresh fruits and vegetables, and though small in stature, seemingly ate his weight in them while there at the dam. He was amazed to find them so readily available at the nearby ranches. He dined with us several times during his stay. He was seeking help for the planned construction of the Ramapadsagar Dam on the Godivari River in his native Madras State in India. He expressed his regret that I declined his offer for assignment on that project, and was rather displeased when Mr. Banks refused to *direct* me to accept his offer. "Well," he said, "that is something that you will have to pursue with Mr. Downs!" Dr. Rao later became the commissioner of Reclamation for India with headquarters in New Delhi.

Then there were groups that gathered in private homes for friendly games of poker. Ours was a penny-ante, dime-limit game, with the last round starting at 10 p.m. There were usually six or seven from the group of Bill Clagett, "Maj." Butler, Carl Cramer, Norm Holmdahl, Don McGregor, Carl Nielsen, Doug Seeley, Charlie Simmons, Joe Turner and me. One evening, after we learned that Doug Seeley had been called up as a reserve officer as the start of hostilities of World War II approached, we decided a parting gift would be appropriate. We arranged to have Clagett stack the deck on the last hand so that Doug, sitting to Clagett's left, would be dealt three aces. All the rest of us would get good starting hands, too. As planned, Doug was to call for two cards and get the last ace, and the others also would improve their hands. As expected, the bidding was more lively than usual, and when the bets stopped for the draw, Doug got a smug look on his face and said, "I'll play these!" We had been outplayed, but he won the pot—a nice token of appreciation we thought. Doug always played them "close to his vest."

Fishing was a popular activity, as was hunting for upland birds and deer in the fall, and poaching was not unknown.

Baseball teams were organized and intertown and other groups batted the ball for the laurels and the applause of the fans. There was tennis and volleyball and swimming in the new pool for a short while, which was heated electrically. The pool had been constructed on an unconsolidated fill and when filled, soon began to crack and then failed miserably when the now-saturated clay and earth dropped from its support. It was rebuilt, however—and inspected—and is still enjoyed by the people of the town.

There were social events and dances, too, both ballroom dances and square dances. Mr. "Ollie" Hartman operated the theaters in Grand Coulee and later in Mason City. Jogging, aerobic dancing, weight lifting, and other 1986 fads for body toning had not yet hit the front pages, and weren't really needed. People *worked* to stay in shape!

And there were babies, many born in the hospital in Mason City. Neither exists today, the hospital was razed and Mason City was renamed Coulee Dam. It is the only town in the state which lies in three counties.

Only once, however, did a new arrival cause this much excitement. Harold Sheerer burned rubber bringing his station wagon to a screeching, dusty halt there at the field office, jumped out and ran over, coat tails flying (he always wore a duster) to slam open the door and shout above the noise of the job, "Bud Burke has twin boys!" It was January 16, 1935. Doug was number one and Don was number two.

Card parties, especially bridge and canasta, were popular and C.D. and Vera Newland's Green Hut Restaurant was the scene of large bridge parties in the afternoons. These events were so frequent that the Hut was often referred to as the "Ladies Pool Hall."

There were churches, too, with myriad activities. The Red Cross and the Boy Scouts and the various school activities were also well supported.

Major and minor events were brought to the people by the "Saturday Evening Post," Bob Ross's *The Star*, Sid Jackson's *Grand Coulee News* (Issue 1, Volume 1, November 3, 1933) and the *Wenatchee Daily World*, which had a young cub reporter—Hu Blonk—on the site of the work full time. He was a buddy of some of the wags and funsters working there, so some of his byline stories were as entertaining as the cartoons. Hu eventually got a bit "gun-shy" about tips on stories and began to check other sources. Once when he heard a rumor that the Bureau was going to build an "ice dam" he knew the boys were trying to pull his leg again. So he went over to the concrete laboratory to verify the tip. His contacts there knew nothing about any absurd thing like that, so he sat smugly thinking he had "had his day." He read of the plan, though, in the *Spokesman Review* two days later He had been *had*—and scooped at that! He still is provoked at himself about it. He later became the *World's* managing editor.

Mrs. Banks was a remarkable lady! Dedicated to Mr. Banks and his work, she devoted much energy as the hostess of the project, entertaining and serving meals to visiting presidents and others in all walks of life. She was a devout member of her church, an active participant in the American Red Cross, and she enjoyed the "movies of the day." She was very friendly; we all loved her and enjoyed the frequent opportunities to chat with her as neighbors. She would often drop in walking home from the Red Cross, her church or the movies over in Mason City. She was fun, and enjoyed friendly repartee—she could dish it out and she could take it—with a laugh!

She stopped in to see us one evening after seeing a movie in which Gypsy Rose Lee had the lead as a "stripper." And I don't mean one of those carpenters down on the job removing forms! She invited Margaret and me out on our front lawn where, with her shoes removed, she pantomimed some of the antics of "Gypsy." She was so clever! She was known as Cupid's little helper, too. At a social function where Guy F. Atkinson, then a widower, was present, he mentioned to her that he would give her $1000 if she could find him a suitable wife. Quickly she responded, "You are looking too far away! You should look closer to home." She answered his next question with, "Rachel Whitehouse!" Mrs. Banks was bubbling with happiness for those two when she told

Theodora Banks, wife of construction engineer, Frank A. Banks. They resided at the dam from 1935 until Mr. Banks died in 1957. May 12, 1945.

Margaret and me of their marriage. "They are both wonderful people and will be so much happier in their years ahead."

I recall only one occasion though, when she was at a momentary loss for words. Nobody in the camp locked their doors in those days, and one night when Wendell Mulkey was sitting in front of the radio listening to the "Richfield Reporter," a nightly 10 p.m. camp ritual for the news, the front door swung open and there was Mrs. Banks, returning from a movie. She had seen the light at Mulkey's, so just dropped in. Wendell was in his pajamas and barefooted—not exactly ready for company. As he jumped to his feet, she was kind of surprised, too, I reckon. Very ladylike, however, she just caught her breath, eyed him head to toe and remarked, "My! What clean, white feet!"

One Halloween Mrs. Banks talked Mrs. Miner into joining her in going to some of the homes in the camp for "tricks or treats," wearing masks and costumes for the occasion—hers was a Pinocchio creation. When they rang the bell at the Turner's, Peggy, a Denver socialite, met their shouts of, "Trick or Treat!" with a cool stare and the comment, "Aren't you girls a bit large to be so engaged?"

My wife, Margaret, has never forgiven me for a remark I made to Mrs. Banks at her dinner table. Margaret was not in camp at that time, and I had been invited to have dinner with Mr. and Mrs. Banks—just the three of us. It was wartime and things were rationed. Mrs. Banks remarked how difficult it was for her to manage the meals with extra guests occasionally, without a source of additional ration stamps. As a sort of confession, she said it had gotten to the sad state that she was now using margarine to augment the short supply of butter, and that she always served a pat of each and so far no one had ever been able to tell the difference. "Can you?" she asked.

Pointing, I said immediately, "That is the butter!"

"Well, how can you tell!" she asked.

I replied with a chuckle, "Why, it is the smallest piece!"

She arose from her chair and came around and pounded on my shoulders with both fists. All three of us enjoyed a good laugh.

Mr. Banks, too, enjoyed a bit of banter. On an occasion when Barry Dibble, a consulting electrical engineer and former Bureau-ite was being teased about his large family—was it *ten children?*—someone of the group asked, "Barry, do you suppose your playing around with all that high voltage accounts for your numerous offspring?"

After the laughter had subsided, Mr. Banks settled the matter, "I think rather it might be the high frequency!"

Besides the "greatest show on earth," the ongoing construction of the dam, we were privileged to see two other performances that live brightly in our memories. The dam was far from the city, and live entertainment was amateurish for the most part. But one Christmas, CBI brought to Mead Circle, in the heart of Mason City, a Christmas scene of remarkable beauty. The manikins were life-sized and they sang beautiful recorded songs and then marched off the stage, still singing. It was beautiful, and I can see it still!

The other memorable performance was completely unplanned, though perhaps it could have been anticipated. A group of us were enjoying a dinner party when our chatter was intruded by the deep-throated hum, then rumble, then high-pitched scream of a big transformer in its death throes and then the snap of the pyroelectrics as the power lines of the 110,000-volt transmission line over in Mason City arced over. The arcs raced from pole to pole, with flames blazing, and the copper conductors red hot and sagging, then cooling and rising again and again as the system would not "let go of the incoming energy." How long it lasted I do not know—but that was 48 years ago

"Robots" as Christmas carolers entertain in Mead Park in Mason City. December 28, 1938.

Eighth grade graduates, Coulee Dam Elementary School, Class of 1942. Standing: Barbara Diamond, Marilyn O'Malley, Lois Wilson, Coralee Murray, A.H. Irwin, Principal, Joyce Mitchell, Gladys Patch, Barbara Reeder, Virginia Stiles, Ileen Richards, Patricia Robbins, Jean Hayball, June Hayball. Kneeling; Vail Luce, Billy Cowals, Charles Hall, Gene Steveson, David Holloman, Neil Sears, Gordon Steveson, Don Cutting, Billy Pontsler, Gene Davis, Clyde Thornton, Bill Miller, Carl Eidelman. May 28, 1942.

and it is still clear in my eyes and ears of memory. I understood that there was insufficient circuit breaker capacity in the transmission system to disconnect the major fault from the system and that the Washington Water Power Company finally had to shut off the water to the turbines at its Lake Chelan power plant to stop the show. But it was dramatic while it lasted!

Mr. Banks was from Maine and he enjoyed the water. He bought a 38-foot Matthews cruiser—christened it *Julie Ann*, honoring his granddaughter. It was an escape from the burdens of the dam and a great pleasure to him and his family and their guests.

SCHOOLS

Bureau specifications for the work typically required the contractors to make all necessary arrangements with the proper school authorities to provide school facilities and instruction, up to and including the twelfth grade, for the families of the contractors' employees living in or near his constructions camps. Such facilities and instruction was furnished without charge to the recipients and the

contractor received no direct reimbursement from the Bureau for such. It was considered a part of "the cost of doing business." MWAK and CBI both supported the schools which they constructed and Mason City High School became an active participant in local events. The Bureau also constructed a grade school building in the government camp, which was first used as the project office while the administration building was under construction. It was converted and used as a school building until the two towns were combined as Coulee Dam. Joe Stansfield was the principal, then superintendent of schools in Mason City for 18 years.

SECURITY

Bureau specifications also required the contractors to be responsible for maintaining good order in the camps and on the work, and to that end required the employment of necessary officers and watchmen to exclude unauthorized persons from the camps and the work. The contractors' constructed quarters and maintained buildings for the detachment of the Washington State Patrol in Mason City

C.C. Beery, local head of the N.R.S. office standing on right. Helen Thomas at his side. March 1934.

The CBI time shack. Employees were all identified by numbered "brass" badges which were issued and returned here. July 28, 1938.

throughout the MWAK and CBI contracts. Also, "guard shacks" at the entrances to the work were established to control access.

When those contracts were completed, the Bureau established a Federal Guard force to patrol the site and to control access to the facilities. That force is still maintained for the protection of the works and the public.

While security was one of my burdens while assistant project manager, the Bureau regional director and members of his staff arranged a meeting at the dam to "iron out procedures" that would govern the relationships and the method of operation while the contractor for the Third Power Plant was working within the dam and the operating power plants. The director stated his requirements to the construction engineer for the Third Power Plant that the personnel working on it would be required to be fingerprinted and issued a badge upon entering the operating plant and dam. The badge would carry the photograph and number of the individual, as was used by all others entering the works without escort.

When the work began, certain workmen from the Third Power Plant activity sought entrance into the dam to "see what had to be

done" and since they had no badges, the Federal Guard denied them access. Immediately Roscoe Granger, the construction engineer, telephoned the project manager at Ephrata, loudly protesting the requirement for badges. He had made no protest when the arrangement was set up by the regional director, however! As a temporary measure then, a guard was assigned to accompany the unbadged personnel into the dam. The matter was promptly pursued though, and the regional director withdrew his requirements. Thereafter, entrance could be gained into the dam and power plant by persons working on the Third Power Plant wearing white hard hats with one or more short stripes thereon. When the regional director explained his retreat on the badge issue, he stated that, "Many good craftsmen in construction have 'records' and don't want to be photographed and fingerprinted." That, I thought, was a most ridiculous reason for permitting anyone entrance to "guarded facilities!" Fortunately, no incidents occurred threatening the structures from "unidentified" persons. I understand, however, that severe damage to the stator of one of the large generators in the Third Powerhouse did occur—caused by parties *yet* unknown.

By 1986 standards, the construction of Grand Coulee Dam was a highly labor-intensive job with many tasks performed with only simple tools. Wages were the minimal specified under the MWAK contract and only two classifications prevailed; unskilled at 50¢ per hour, and skilled at $1.20. At the outset, the labor force was not organized, though some of the workmen belonged to craft unions. But with time, organizing efforts gained ground and on July 30, 1937 MWAK signed a closed shop agreement with the A.F.L., American Federation of Labor, whereupon the C.I.O., Committee for Industrial Organization, filed a protest. The N.L.R.B., National Labor Relations Board, set up a hearing for October 18, 1937, which was later postponed, and in fact, never held. The C.I.O. retreated from the scene. The job was winding down, so no effort seemed to appear to displace the non-union workers with those carrying a card. Nor was there then any change in wages nor working conditions. Shortly after the bid call for completion of the dam, rumor on the job carried the word that the Kaiser Co. had reached "an agreement with labor" that provided for increased wages and no work stoppages in the event that the company was the successful bidder. The specifications No. 757 for completion of the dam and power plant included 123 classifications for work and set the minimum wages that would be permissible under the contract. The rates per hour were set as the prevailing rates in the locality and ranged from labor, unskilled 60¢ (of which there were few); laborer, concrete 65¢; several classes of helpers 75¢; carpenters, drillers, blacksmith, boilermaker, concrete mixer operator, dispatcher, batch plant operator, electrician and others of the crafts, all $1.20; to crane and hoist operators at $1.50.

The rates shown in the specifications indicated that certain of the crafts—particularly the operating engineers, received increased rates of pay while most of the other crafts' rates remained the same as they had been recently under MWAK. Also, certain of the semi-skilled classifications' rates were increased. I never heard how the rates were determined. But, insofar as I can recall, there was labor peace throughout the CBI contract and no strikes.

This agreement between the contractor, CBI and the A.F.L. local and international unions was dated December 9, 1937. It established the rates of pay for the various classifications of labor, banned strikes, provided for arbitration of disputes under certain conditions, placed upon the unions the responsibility for settling jurisdictional disputes, and gave the contractor the sole right to determine the fitness of employees to do the work assigned.

Several of the skilled craftsmen employed by the Bureau were card-carrying union members and many crafts were employed in the installation of the power plant equipment. The Bureau determined the wages and the crafts were very zealous of their jurisdictions. Some disagreements occurred as to work assignments and working conditions. More order was brought into the process when the Bureau negotiated and signed an agreement with the Columbia Basin Trades Council on April 27, 1949. The Council included all of the craft unions employed on the Bureau's non-contract work. When I became assistant manager of the project in 1953, I also became the chairman and spokesman of the Bureau's negotiating committee until my retirement in 1970. My impressions of the work accomplishments under a craft-oriented work force in summary is that the skills of the workmen may be somewhat higher and the work performed at a greater degree of perfection than when the work force crosses craft lines in the work process. And it requires more detailed scheduling and costs are probably

The Bureau (B) executed an agreement of labor (L) policy with the Columbia Basin Trades Council. Front row, Jack Barron (L), V.H. Mills (L), A.F. Darland (B), F.A. Banks (B), Bob Smith (B), H.A. Parker (B); Back row: K.C. Will (L), Lionel Loiseau (L), S.S. Bartels (L), Norman Lull (L), Monica Gray (B), G.S. Hobs (L), R.A. Milne (B), R.C. Miller (B), S.C. Riggle (L). April 27, 1949.

Payday for MWAK crews. The contractor was required to pay each workman in *full* each week. Pay was made in two checks, one in the amount of any debits for quarters and subsistence (if any). That check was signed back to the contractor at time of delivery. June 4, 1937.

higher than when craft jurisdictions are less precise. But the agreement did promote more harmony than existed before it came into being—that is my impression from this distance. The agreement merely formalized the organizational structure in existence since the Bureau first started to install the machines in the plants.

EMPLOYMENT

At the outset, the work in preparation for the start of construction of Grand Coulee Dam used funds totalling $377,000 advanced by the State of Washington and was expected to be manned with residents of the state with preference to the adjoining county residents. But some of the funds were available for personnel then working for the Bureau of Reclamation under civil service appointments, so some transfers of Bureau personnel occurred—I was one of these. As I recall, most of the employees on the work force were in a temporary status, and clearance was required from "Democrats," so a number of those employees have told me. They also said that if you could play baseball your chances of getting a job were not lessened. Since the work was somewhat strenuous, persons with athletic abilities perhaps were considered more able to do the physical parts of their assignments—of course, a winning baseball team wasn't something to be ashamed of, either! But the clearances were not viewed by many as onerous or difficult to obtain. Many of the men on those first survey crews—hired as chainmen and rodmen at wages of $100 a month were recent graduates from the state universities' engineering schools, had been outstanding students, and had good minds as well as brawn. They wanted to be part of this great undertaking. Some of them, in fact all of them, moved up to more responsible jobs as soon as such openings occurred or moved to other opportunities. Workers seeking employment with the contractors on the two exploratory contracts by Rumsey & Co. and Lynch Bros. were "cleared" through the local office of the state Columbia Basin Commission. Those two contractors were paid by the state treasurer.

When the project became a federal undertaking, the office for clearing prospective employees became the National Re-employment Service, organized and operated by the Department of Labor. The local office served the portions of Grant, Douglas and Okanogan Counties adjacent the damsite.

Job seekers were many! And they came from all walks of life and many came under assumed identies. Out-of-staters were not given top priority, and Washington residents with experience and the skills needed for the job were few in number. Those who had gained or sharpened their skills on Boulder Dam and who were headed this way had residence problems because of these local and state priorities. When CBI contracted with organized labor at the start of that contract, some told me of having difficulties getting on the referral list because union business agents wouldn't give them a card—unless appropriate "fees" were advanced in cash.

Referrals to all contractors on site numbered 11,671 in 1936; 12,963 in 1937; with the total during 1933-1937 being 42,257. I recall that the peak of employment on the dam was said to be 8,800. That may have included businesses, also.

The following statistics from the Bureau of Reclamation Histories indicate the magnitude of the labor requirement for building the dam and reservoir.

MWAK EMPLOYMENT

1934	487,425	Manhours
1935	5,299,969	Manhours
1936	8,441,374	Manhours
1937	8,694,803	Manhours
1934-1937	22,923,570	Manhours
	$19,502,492	Total
Payroll Average	$.85 per hour	

CBI EMPLOYMENT

1938	4,733,179	Manhours
	$5,135,320 Total	
Payroll Average	$1.08 per hour	

U.S. EMPLOYMENT

1933-1938	4,906,904	Manhours
	$3,726,609 Total	
1939	2,261,023	Manhours
	$1,827,954 Total	

TOTAL EMPLOYMENT
U.S. and all Contractors

1933-1939	37,000,119	Manhours
	$34,650,244 Total	

The Bureau paid by check in the field to hourly workers. Pay days were at the end of the month. Here Alex Harker is delivering checks to Oscar Pryor's crew. Oscar at left. November 1, 1935.

The average work week was reported to be 37.5 hours per week for the contractors' personnel through 1937. Typically, the Bureau employees worked 44 hours per week until March 1, 1942 when the 48-hour week (with pay and time-and-one-half for overtime) came into vogue on the project. After the War, work resumed on a 40-hour-a-week basis.

I have been unable to find the employment statistics for the remaining years of construction.

WOMEN IN THE WORK FORCE

This was a man's world and few women had penetrated its confines. But there were a few brave ones. Cap Beery had three women in his employ at the N.R.S. Dusty Rhoades had a secretary/clerk who kept the books and wrote the letters and reports. Also I believe there was one woman employed by David H. Ryan. When MWAK arrived on the site so did women switchboard operators, secretaries and clerks—but in total they were few in number.

At the outset, no women were employed by the Bureau, but then

Sub-committeemen, House Committee on Appropriations on an official visit to the project, accompanied by Congressman Hal Holmes of Washington and Assistant Commissioner W.E. Warne of Bureau. Warne, Ben F. Jensen, M.C. Iowa, M.J Kirwin, M.C. Ohio, W.F. Norrell, M.C. Arkansas, Hal Holmes, M.C. Washington, Frank A. Banks, Construction Engineer, describes features of one of the 108,000-kv generator stators. September 6, 1943.

along came Miss Helen M. Newell who was hired as a clerk-stenographer. Later she went off "to be more help in the War effort," and, being slight of stature, did what most could not—she was a riveter in the wings of airplanes as they were being assembled. Later she became an author. *Hard Hats* is her version of how the dam was built.

The Bureau did not provide quarters for single women, but MWAK did when it built the "women's dorm" as part of the contractor's camp in Mason City. Then more women gradually appeared in the offices and while there was a scarcity of manpower during the War, women came into the work force down on the job as timekeepers, toolroom clerks, materials checkers, elevator operators, taxi or shuttle drivers, equipment dispatchers, gardeners, etc. And they were effective! Mary Jane Rhoades, Kathy Patch, Wanda Wambsgans and Marie Hayball were some who come to mind.

Smoking by the "fairer sex" was taboo at the work stations, however, so the ladies lounge became the smoke shop. Mr. Funk kept a log of time so spent, (loitering was not a part of his work ethic.) When Sophie was absent from her desk for 45 minutes one day, he faced her at her desk. "You were away from your desk for 45 minutes!"

Congressman Henry M. Jackson and Senator Hugh B. Mitchell visit the dam. September 9, 1945.

General Wainright visits with Mr. Banks at a luncheon at the coffee shop. November 13, 1945.

Presidential candidate Governor Thomas E. Dewey signs the register at the visitors center. February 11, 1948.

Charles "Charley" Osborne lived at the damsite for almost 70 years. October 13, 1948.

"Was I?" she replied. "I thought it was longer than that!" No one had the upper hand with Sophie.

After the War the question of women smoking at their desks arose and I remember Bert Hall's decision at our field office staff meeting. "Well, okay. But I don't want any women at my reception desk banging away on a typewriter with a cigar dangling out of the corner of her mouth!"

Mention of women in the work force brings to mind Doug Wood's sage observation, "Woman abhors the sight of a stationary male!"

MINORITIES IN THE WORK FORCE

In 1933 black people were not numerous in the counties adjacent the damsite if indeed there were any, and the few within the state of Washington lived in the coastal cities principally. Also, few black men were employed in craft skills at that time when the dam was started. The native Americans on the adjacent Colville Reservation were ranchers or loggers or worked in the sawmills as their main lines of employment. So, it was a number of years before I noticed some black men and a few native Americans in the crews working on the dam. As I look back to those years, the impression I have is that "minority" workers were usually employed in unskilled tasks.

Landy Harrell was a black man I knew, who was employed by the Bureau in the government garage in Coulee Dam servicing cars and doing related tasks. Later he went into business in Grand Coulee and was a city councilman there for a dozen years.

There were a number of Indian workers who were employed by the contractors as well as the Bureau. I can remember a few; Whitlaw, Eddie Nanpooyah, Yellow Wolf and Pete Moses. There were others, but these I knew personally. After the War, many black families came to the dam seeking work. They had been in the shipyards and at the Hanford works of the AEC. Many were short of funds and the Coulee Dam Federal Credit Union made arrangements with Mr. Funk, the Bureau chief clerk, whereby loans made "for living expenses till payday" would be repaid when the borrower came to the credit union to pick up his check. Some of these men had never been advanced money before—except from loan sharks. They were very appreciative—some had to borrow several times during the first few pay periods. I know the credit union helped many of the Bureau employees, both black and white; I was one of its charter members and served on the supervising committee.

VISITORS

Grand Coulee Dam was the "Eighth Wonder of the World" in the making and it brought visitors from all walks of life to see it as it took shape. There were presidents: Franklin D. Roosevelt and Harry S. Truman; presidential candidates: Governor Thomas E. Dewey of New York; royalty: Prince Olaf of Norway; emperors: Hailie Selassie of Ethiopia; cabinet members: Harold L. Ickes, James A. Farley, George H. Dern and David C. Roper; senators: C.C. Dill, Hugh Mitchell, Warren G. Magnusson of Washington, Norris of Nebraska, Dewart and Wallace of Montana; representatives: Henry M. Jackson (later senator), Hal Holmes and Walt Horan of Washington for example; governors: Clarence Martin, Arthur Langlie and Mon C. Wallgren of Washington; generals: Jonathan M. Wainright (hero of Corregidor); scientists and engineers both U.S. and foreign from China, France, India, Japan, Russia and others; government officials, U.S., state and

(Continued on page 55)

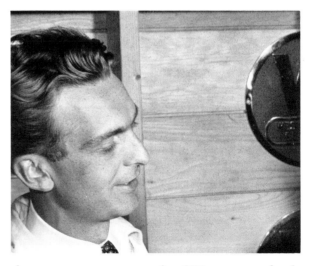

Chet Huntley, a young announcer for CBS, came to the dam for the announcement of the beginning of commercial power production from the main units of the Grand Coulee Powerplant. October 4, 1941.

The little "tourist train" gave the visitors a free ride and a good view inside the Left (west) Powerplant—and avoided the traffic congestion and interference with the work. May 4, 1947.

A Junior League delegation show their interest in the dam. May 9, 1940.

One of the many visiting crowds that come to see the dam. September 7, 1936.

President Roosevelt was very comfortable and at ease with the crowd of well-wishers. October 2, 1937.

Governor Clarence Martin came over from Olympia to dump the official "first concrete" in Grand Coulee Dam. December 6, 1935.

Governor Clarence Martin in cap of locomotive engineer at official transfer of U.S. Construction Railroad to MWAK. Also Francis Donaldson, MWAK chief engineer, J.E. Kinman, Spokane Chamber of Commerce, Jim O'Sullivan, Columbia Basin Commission, E.F. Blaine, Guy F. Atkinson, Vice President MWAK, Ben H. Kizer, Chairman, Washington State Planning Commission, Wm. M. Clapp, E.L. Kier, Vice President MWAK, Harvey Slocum, General Superintendent, MWAK. July 29, 1935.

local; the Junior League; authors: Louis Adamic; the press and radio: Chet Huntley; and "the public" from across the U.S.A. and from foreign lands. They came in such numbers that special facilities, though not elaborate, were provided for their comfort and safety, to let them see from the vantage of "vista houses," where they could hear descriptions of what was going on literally before their very eyes. They rode down to the left power house on the "tourist train," a flat car with seats pulled by a little diesel locomotive. So, they went away believers! Perhaps they didn't really comprehend the size or the significance of the dam and the power plant and those huge machines. Perhaps they didn't really believe all that Charlie Osborn or John Holland told them there at the vista house. But they left in awe.

The real dignitaries of the time, of course, were escorted by Mr. Banks, and he and Mrs. Banks hosted many of them at their table—without an expense account.

The great, white sheet of clear water flowing over the spillway of Grand Coulee Dam is extraordinarily beautiful when viewed either day or night. But certain interests in the state began to urge that the spillway be bathed in colored light at night for the benefit of visitors. The idea soon gained support at the congressional level and a bank of colored lights was installed as a feature of the project. It has proved to be popular and continues as a seasonal evening attraction. Now, however, the discharge of water over the spillway uses water from storage which would otherwise be utilized in the generation of power, so lighting displays are restricted.

DEDICATIONS AND CELEBRATIONS

Dedications and celebrations are bonanzas for politiciians and the media and there were a few nuggets for them in the past five decades of the dam. First, there was the "first spadeful of earth," turned at the future site of the dam on July 16, 1933. The next year President Franklin D. Roosevelt came to the site on August 4, 1934 and spoke to a large, enthusiastic crowd who had come from the "far corners." He returned on October 2, 1937 to review the progress. Governor Clarence Martin came from Olympia on December 8, 1934 to drive the golden spike when the railroad to Grand Coulee was laid and he revisited on July 29, 1935, donning the cap of a locomotive engineer to ride alongside Bert Denny at the throttle, I believe, when the railroad to the dam was officially turned over to the contractor MWAK. He returned on December 6, 1935 to dump and vibrate the "first official" bucket of concrete in the dam. I noticed, however, that he failed to wear the clothes and hard hat of a concrete hand. The concrete had actually been dumped onto concrete which had been placed earlier. The first was put down on the bedrock on Thanksgiving Day, November 28, 1935. Martin made it official, though, for the press cameras witnessing the event.

The highway bridge over the Columbia River was opened to traffic on January 27, 1936.

On March 22, 1941 the first power generated at the dam was connected into the Bonneville lines. And on October 4, 1941 the first of the main units began producing power for industry and the homes in the Northwest.

Mr. Banks took a chain saw and felled the last tree in the clearing of the reservoir on July 15, 1941. The contractor, CBI completed all work on the dam under its contract on December 31, 1941, and turned the *dam* over to the U.S. Bureau of Reclamation the following day, New Years Day, 1942.

By early June 1942, Lake Roosevelt had reached sufficient height

President Harry S. Truman dedicates Grand Coulee Dam and Franklin D. Roosevelt Lake. Construction Engineer Frank A. Banks to left of podium, Senator Warren G. Magnusson, Regional Director Harold T. Nelson and Commissioner Michael L. Strauss of the U.S. Bureau of Reclamation to the right of President Truman. May 11, 1950.

Initial flow of water from Lake Roosevelt over the spillway crest after completion of Grand Coulee Dam. Construction still in progress on Right Powerhouse. June 2, 1942.

Forty-eight young ladies pour ceremonial water from the 48 states at the beginning of water delivery from Grand Coulee Dam for the irrigation of the Columbia Basin Project lands. May 7, 1951.

for the initial spill over the 1,635-foot-long spillway crest. The reservoir continued to rise until it filled to elevation 1,290 on July 16, 1942. Gates were operated thereafter, so that the level was not exceeded to avoid backwater rising into Canada.

President Harry S. Truman came on May 11, 1950 to dedicate Grand Coulee Dam and name the reservoir Franklin D. Roosevelt Lake. The last of the main units went on the line, completing the power plant installations with an installed, rated capacity of 1,969,000 kw on September 14, 1951.

That great milestone was overshadowed powerfully with the real "hoopla" of May 7, 1951, when the pumps were started and water for the irrigation of project lands began filling the equalizing reservoir. To indicate the participation of all states, jugs of water had been solicited from each, and the ceremonial jugs were spilled into the feeder canal by 48 fair young maidens in white. The equalizing reservoir in the Grand Coulee was dedicated to the memory of Mr. Frank A. Banks, the "boss" in the construction of the dam and appurtenances and the beautiful body of water was named "Banks Lake" on September 5,

1958, in honor of that great man. Mr. Banks died on December 14, 1957. His memory is cherished. His ashes, I suspect, are resting at the bottom of the great reservoir behind the dam.

Lastly, on the occasion of the 50th anniversary of the symbolic "first turning of earth," the mighty Third Power Plant was dedicated on July 16, 1983. That was a unique and memorable day—not from the utterances of the politicians who mounted the platform, to recite or to remake the history of the dam with scant recognition given to those who *built* the Third Power Plant or the dam—but memorable to hundreds of workers who returned to the site of their labors. They didn't come to hear speeches, but to relive with their associates those days long gone—and felt the satisfaction of having been "part of the act."

These dedications have been important social, commercial and political events over the years, both to the public and to the media, but they pale in significance when viewed in the light of the dedication to the job in the hearts of those who were on the job! They were a part of it!

An assemblage of part of the Bureau force at the administration building. Ed Argesinger, Snow (?), Wendell Swanson, _____, _____, _____, Bill Moore, Jack Cavin, Steve Girard, _____, Fred Sharkey, A. Francis Meyers, E.J. Niemen, W.A. Allison, Chuck Weil, A.F. Darland, George Teufel, E.M. Burley, C.M. Cole, Tom Ostliffe, T.R. Anderson, R.W. Smith, Francis "Maj" Butler, E.R. Hogan, Jack Sowle, Glen Burrows, Harold Sheerer, Bob Rockmanov, Carl Buckholtz, John Emig, _____, Harold Tennant, George I. Owens, Axelson, Oscar Dike. March 8, 1936.

Chapter III
MANAGEMENT AND RESPONSIBILITIES

MANAGEMENT

I recall seeing the *Manual of Reclamation Laws, Rules and Regulations* and an employment card. The manual, when I arrived at Almira to work on the dam, was a single volume about an inch thick. And the employment card stated the basic policies—including the committment of the employee to work overtime, which would "be cheerfully given," and the government's offer to furnish "rubber boots and slickers." The Bureau employed well-qualified engineers and construction men and gave them a "loose rope," expecting them to do the job assigned to them; to report progress and problems that they couldn't handle or which needed the participation of "greater expertise." In the field, the lines of authority were easily apparent. But at the outset, at least until Bert Hall arrived on the scene, there was a tendency to make field decisions at the office. I recall in December 1934 a contractor told me that Mr. Banks in Almira had authorized him to stop drilling a well in one of the slide areas, based on what the contractor had told of the conditions in the field. The contractor, however, didn't have the facts of the situation. I thought the decision was based on an error, so telephoned Mr. Banks and told him what the conditions were. I recommended that the drilling continue, whereupon he replied, "Well, I guess I had better leave this in your hands."

When Bert Hall came to the field office on his first day at the dam, he and I were reviewing the current work and staff when the phone rang. Bert picked it up and I heard both ends of the conversation. "Mr. Hall, this is Ryan. I have a problem up here on my job and Mr. Darland referred me to you."

"Take the problem up with the inspector; you will find him there on the job. His name is Boggess." Mr. Banks had great confidence in Bert Hall and gave him a loose rope, too. As far as I know, it was never jerked or shortened, either.

Mr. Banks and others at the project office, kept informed by daily reports prepared by Pat O'Malley and by staff meetings in which he stayed until his concern-of-the-moment was answered. The chief engineer, too, was kept informed by frequent reports and later, when coordination of design, procurement and construction progress became so critical, Mr. Frank B. Cook was assigned from that office to spend alternate weeks in the chief's office and at the dam as a coordinator. Through Frank's attention, the grease got on the right wheel at the right time.

After the War—after the old-time Commissioners of Reclamation, Dr. Meade, John Page and Harry Bashore had retired—a new breed of commissioners were introduced. They, for the most part, had little knowledge of the engineering work the Bureau did, but had a lot of influence. They were politically acquainted and oriented and undertook to greatly expand the Reclamation program. Elaborate work schedules were set up for each project, organizational structure was standardized, administration and development became an important part of the work scene. Purchasing, supply, accounting and finance departments ballooned and overhead costs did not diminish. The *Manual* grew so that it would not fit on a shelf which could

Bert A. Hall, chief inspector on construction of Grand Coulee Dam. Later was supervising engineer of civil and structural activities. Born July 2, 1893, died April 24, 1990. October 10, 1945.

accommodate the *Harvard Classics*. Keeping the manual up to date became burdensome. Inter-office correspondence multiplied and that was before Xerox. Paperwork was such a big part of the work that at a construction engineer's conference in Denver, I mentioned that we now had a new measure of accomplishment of work—"A pound of paper for every yard of muck."

ORGANIZATION AND STAFFING — Bureau of Reclamation

In an earlier paragraph mention was made of the initial field staffing of the Bureau as the preliminaries got underway in 1933 and 1934 and the first four months of 1935. With Bert Hall's arrival on the job, the activities of the Bureau personnel were changed somewhat. The engineering office was unchanged—drawings, correspondence, computations for mapping and delegated designs and for payment of contractors' earnings, preparation of correspondence and contracts for appraisal and acquisition of land and land rights, preparation of correspondence with contractors and the chief engineer who was the contracting officer for the contracts. Payrolls, personnel, purchasing, accounting, rentals, and receipt of government-furnished materials were all the responsibility of the chief clerk.

Some new faces appeared at the office, and although I'm not sure of the order in which they arrived. They included Carl Nielsen, Herman Kelsch, Joel Johansen, Bill Cree, Bob Oyler, Don Anderson, Jim Flynn, Tom Jose, Bill Hay, Bob Culp, Charlie Zack, Carl Shaw, Bob Garen, Charles Hoag, Winifred Williams and Monica Gray.

The so-called field activities were all under the supervision of Mr. A.F. Darland, the field engineer.

It was Bert Hall's responsibility to accept the contractors' work that was built strictly in accordance with the contract, the specifications and the change orders and orders for extra work. All these were assured through his inspection procedures and reports thereon.

The inspection of the various contractors' concrete manufacture, operating a testing laboratory in that connection, and related duties were Mr. O.G. Patch's responsibility. Grant Gordon had the burden of soil investigation, diamond drilling, test pitting, geological matters and later efforts to control sliding by drainage of the riverbanks, and the investigation of the damsites for the balancing reservoir (Banks Lake). As I recall, Grant was the person promoting and supervising the construction of the Ice Dam to stabilize the toe of the sliding ground. But I think Lloyd Froage first suggested it. Operation of camp services, maintenance of buildings, utilities, garage, fire protection, and miscellaneous minor construction by Bureau workmen was directed by Wendell Rice. R.T. Sinex planned and directed the finish grading and landscaping of the entire campsite. Later this job was handled by Bob Yerxa. C.M. Cole supervised all surveying. First order leveling and triangulation and establishing control points throughout the project was performed by Carl Berry and his crews. Althe J. Thomas' party surveyed the needed lines and grades in the camp and for the highway bridge. Later he was the locating and construction engineer who was responsible for the relocation of roads and railroads in the reservoir area with headquarters at Colville. At the dam, E.J. "Jake" Niemen was responsible for the surveying on the west side of the river and Orville I. Craft and later A.A. Brownson, both from Boulder Dam, surveyed the east side. Later C.E. Burningham was in charge of surveys at the dam for two years or so. Fred Berry moved to Ephrata for surveys on the irrigation phase of the project. Harold Sheerer, who later became an

Orin G. Patch, chief of concrete control and testing activity June 1936.

inspector, was a party chief on the construction of the railroad. Other party chiefs I recall were: Paul Piper, Ray Smith, Seth Norris, Ross K. Tiffany Jr. who also worked as an inspector, Tom Wall who was also from Boulder, A.A. Brownson, Luther E. Cliffe, Carl C. Scott, Jerry Cranford, Bill Cowals, Tex Walker, Gordon Bockemohle, and others. Some of their crew members included: Chuck Prahl, Don McLaren, Hal Newton, Wendell Clark, Alister McNab, Ableson, C.A.J. Pauw, George Neff, Trig Hansen, Ellis, Snow, Phil Young, Fred McCune and Hydorn. Each individual had an important part in keeping everything lined up, on grade and plumb. They measured all quantities for payment. Ever since Mr. Banks, with that first assembly of the survey chiefs as his party put his heel to the ground high up on the west abutment axis of the dam and said, "Put it there," the surveyors continued to control the grade and alignment. They not only got it in the right place, they had it pointing in the right direction.

There were a number of changes in the Bureau organization over the years, but the team that Mr. Banks put together was capable, industrious and dependable and self-reliant. By 1945 though, the top management was thinned out. I recall the different roles that had to be absorbed when Mr. Banks was assigned the task of Regional Director when the Bureau was divided into seven regions with the Northwest being region 1, with headquarters in Boise, Idaho. Earlier, on May 4, 1939 while Supervising Engineer he had been assigned the added duty of Acting Administrator of the Bonneville Power Administration, with headquarters in Portland, Oregon.

But back at the old "Charley Osborne Ranch" there were some drastic changes. Mr. Miner had died on May 31, 1943; Mr. Darland had resigned on May 15, 1945 and moved to Tacoma. He was later to return, however. Fred Sharkey had had a serious heart attack. C.F. Christensen had retired and his replacement, A.P. Newberry, chief of the electrical and mechanical division, had resigned to go into private business. So Bert Hall's burdens of responsibility were greatly magnified. The job rolled right along, though, but I doubt if Bert ever got the recognition he deserved for the performance demanded of him at that time.

With the increased activity in the irrigation division of the project, Mr. Banks resigned on November 1, 1945 as Regional Director and secured Mr. R.B. Williams on the dam project as his principal assistant on the 18th of that same month. His title was Assistant Supervising Engineer. Both highly capable, they were sensitive to the efficient running of an organization and had a way with people. I recall one instance as an illustration; the Bureau forces were dredging the river downstream of the spillway bucket and the work had a high priority, but the fall deer hunting season was approaching and many of the riggers and dredge operators wanted time off for those annual rites. Mr. Darland had decided that the crews on that project couldn't be spared and had about settled the matter when Mr. Williams rolled his eyes up to the ceiling and suggested that denying those men their right to go hunting would result in lowered morale. It was an efficient crew, he pointed out, and it would not be in the interest of good production to keep them at their jobs. Besides, they all had annual leave coming—it had to be granted sometime. "Let's shut down the dredging for one week during the first week of deer hunting." Well, that settled the matter!

In 1933, while I was still in Denver, and thinking about a transfer to the field, I inquired about the various jobs among the far-flung activities in which the Bureau was involved. Oscar Rice and others who had been on other construction jobs with Frank Banks and R.B. Williams vowed they were "top of the line." How fortunate I was to work with both of them—and on the same dam!

R.B. Williams retires. Mr. Williams, Assistant District Manager, retired from the Bureau on October 14, 1949, and was congratulated by district manager F.A. Banks for his long, successful service with the government.

At the outset, an accurate baseline 2,000 feet in length was monumented on the ground on the flat land lying above the right bank of the river. Supports along the staked line were used to hold the calibrated tape on a uniform slope and in accurate alignment while measurements were made at observed temperature and tape tension. When the exact distance between the baseline monuments was calculated, fixed monuments were established about the canyon walls and by triangulation methods and computations the locations of the points were determined. From this initial baseline and triangulation points, a grid system was established using lines parallel to the axis of the dam and perpendicular thereto for controlling the building of the structure. The grid lines were identified on the ground by fixed visible and undisturbed targets on the canyon walls. By transit then, the

Althe J. Thomas, C.M. Cole and construction engineer, Frank A. Banks set the axis of Grand Coulee Dam. September 9, 1933.

location of the lines down on the work areas could be fixed readily for control of placement of forms and fixtures.

The U.S. Coast and Geodedic Survey monuments in Eastern Washington provided point of origin for determining the elevation—above sea level—of control points, or bench marks, both at the dam and throughout the entire project. ''Level surveys'' from such control points were extended onto the job site and used in setting the elevation of the works constructed.

Vertical alignment of features was provided by use of transits and by plumb lines. The surveying and mapping equipment used for the first generation of work on the project was such as had been typically available for a number of years. Transits were used primarily to fix horizontal alignment and levels for elevations. Measured distances were by tape or by calculation by triangulation. Mapping was accomplished in the field using plane table and alidades. Radar, electronic measurements, accurate aerial mapping and hand-held computers were still for the future.

Mention of hand-held computers, though, brings to mind the progress that has been made in computer science and equipment. When I started my career, the computers being used in the chief engineer's office were typically electrically-powered devices, or hand-cranked, but the movable carriage was cranked along by hand. Ten rows of keys numbered 0 to 9 were punched as needed, which placed the multiplicand in the keyboard, then multiplication was performed by holding down the + key, which activated the motor. Multiplication was actually an addition function then; the carriage moving along to each integer of the multiple in turn to repeat the process until the product was shown on a lower set of dials in the carriage. The multiplier appeared in the upper set of dials. Visual comparison was then made of the keyboard and the upper dials to verify the correctness of the input before recording the product. Division was performed similarly by punching the – key. In those days—1931—we had only one such machine with 13 rows of keys. It was said to be one of a kind! Within just a few months, however, desk-type calculators with power-shifting carriages came into use and shortly thereafter, automatic ones where multiplication was accomplished by entering the multiplier and the multiplicand into the dials, then pushing the × key or the ÷ key. Now, the little, hand-held calculators will perform in minutes computations that took us hours!

Employees on the project doing surveying were identified as "engineers," while personnel doing quality control were "inspectors," regardless of professional skills.

On May 1, 1935, Mr. Bert A. Hall arrived on the job and all inspection of the construction work, other than concrete control activities then became his responsibility. He was of the "Reclamation family;" having started his career on the Minidoka Project at the age of 15. He moved along to the American Falls Dam, then to Owyhee and later Boulder/Hoover Dams. He took charge! He had answers! Things were done his way! He was unwavering and courageous and when a decision was made, you could count on it to stand "tomorrow." He was specific, stern, and fair. He went by the specifications—and looked you in the eye! Bert Hall, more than any other person, was the one who put Quality into the building of the dam. He was a great influence on my career and for 15 years he was my tutor and my mentor. And I was one of his team!

From Boulder he brought experienced engineers and inspectors; Don Walter, Stubby Jackson, G.A. Warning, Dave Kime, Byron "Toots" Boston, Horace Taylor, Albert Moser, Jess Roeschlaub, Walt Chubbuck, Manuel Aaron, Don Brooks, Clyde Wells, Charlie Peck, V.J. Peterson, Walker Wilferth, C.S. Stover, Howard Bowman, Walt Greenwood, Stan Martin, Jim Motsenbocker, A. Smithies (?), Bill Wright, Chet Purdy, Ed Kitchen, Louie Ackerman, Kenny McDonald and Tom Wall. Too, there was W.I. "Bill" Morgan, a master mechanic and erection engineer, an old timer in Reclamation. Bill was a gentleman. He was the effective, knowledgeable chief of the inspectors on the metal work, which included gates, gate seals, bulkheads, drum gates, etc.

Morris Tibbals was the mature chief of the crew which inspected the erection of the highway bridge and the repair of the "tilted pier." He later was the receiving inspector for the government-furnished metal work. L.V. Froage was an experienced refrigeration man for the ice dam. Mr. O.G. Patch had arrived the previous year and he, too recruited experienced engineers and technicians from Boulder Dam—Douglas Wood, Walt Winner, Wendell Mulkey, Earl Moore and C.T. Douglas.

Those from "down below," as they all identified Boulder Dam, knew what a big job was and they were right at home and immediately effective in getting the team working. They were good teachers in training the rest of the team in the matter and manner of quality control of construction. The rest of us owe them all a debt—perhaps still unpaid. Some of those debtors whose names I still recall were: Bill Allison, Cap Arnold, C.E. Benjamin, Earl Banker, Paul Bickford, Olney Boggess, Carl Buckholtz, E.M. Burley, "Maj." F.S. Butler, D.S. Caton, A.L. Colwell, D.S. Davis, George Duecey, O.J. DeSpain, George Grant, Bob Hogan, Carl Jones, Dick Loudon, George Kreiger, Herr, "Hiney" Hopewell, Don Humes, Sam Huston, Hertice Marsh, Willie McKay, Pelkey, Bruce McLean, Don Schlapkohl, Ray Seely, Jack Sowle, Harold Sheerer, George Straalsund, Wendell Swanson, Johnny Waters, Borden Wilbor, Chuck Weil, George Teufel, Norm Holmdahl, C.F. Royce, Leroy Shively, J.P. Elston, J.M. Erwin, M.H. Larson, Dominic Magnetti, A.F. Moore, Fred Powell, Fritz Hegge, Francis Myers, F.C. Utter, R.W. Murphy, L.R. Engvall, M. Kaiser, V.A. Prisadiski, H.O. Roen, Vern Votaw, D.A. Daly, E.K. Strock, A.E. Washburn, J.P. Swann, W.H. Wisniski, Art Jewell, A.R. Sather, E.O. Niemen, Archie Truesdell, Charles Seeholzer, G.V. Alvensleben, R. Delivuk, Emil Gehri, Phil Joray, R.A.W. Krows, Percy Pharr, O.D Young, Cull White, Bob Livingston, Carl Cramer, Jack Neff, Charlie Parsons and C.G. Rossman. C.S. Hale came in the late thirties. He too had worked with Bert Hall on Idaho projects and was very quiet, capable and industrious. The local additions to Mr. Patch's specialty included C. Brim, Art Brunstead, Bill Clagett, Vic Gezelius, "Mar" Hawkins, Ted Mann, Guy Reynolds, John Vertrees, Jack Wells, Wilbur Roush, Henry Roush and Dale McMillan.

THE CHECK OUT

Within the permissive of the specifications the contractors chose which blocks to construct each shift and projected the work plan several shifts ahead. From that plan, work was assigned by the shift boss to the various craft skills required in each block and the order in which the blocks should be started or completed. In any block, for example, first the initial cleanup of the concrete was done; then the forms were raised and braced into position for that "pour," using control points established by the "engineers." Next, the cooling pipe crew, the grout crew, the grout stop welder, the reinforcing steel crew—if needed—would each perform their assigned work within that block, coordinating their work so that their crew would be out of the way when the cleanup started. That crew then removed all objectionable and foreign material and cleaned and washed the concrete surfaces as "spotless" as the kitchen floor. Now the block was ready to receive concrete.

While the final cleanup was in progress, the "engineers" were called by the contractor's okay-man to verify the presence of all required installations and record the quantities of government-furnished materials installed and to check the forms for position. The inspector was also summoned by the okay-man to check out the block for final "Okay to pour." After this routine was established, the work advanced on each shift so that the blocks were ready to receive concrete

and had the final okay signed by the inspector by the time the concrete crew had finished a block and had moved in.

But that did not always happen—unforseen problems such as omissions or damaged materials, plugged grout pipe or cooling pipe, plugged drains, reinforcing steel out of position or not secured and incomplete preparation of the concrete surface are examples where the ideal flow of work was disrupted. And there were numerous instances where the foreman on the job or his shift boss raised objections to the inspector's decision regarding the acceptability of the work and the delay in the final okay. The shift inspector on duty may have been called to review the demands of the inspector with the foreman or shifter and if the impasse continued, the shifter might decide to just forget that block and concentrate his efforts on another and refer the

Frank "Pop" Smith, MWAK assistant superintendant. May 27, 1937.

Mr. and Mrs. Silas Mason were frequent observers of the MWAK operations down on the job during the first year of the contract. Late 1934.

problem and the disagreement to the concrete superintendent who would put the problem to the contractor's "okay-chief" who would then take the problem or dispute to the attention of the Bureau's "okay-chief." There was naturally a certain amount of rivalry among the shifts of the contractor's organization, as each tried to hold the record of the most "okays" or the most concrete placed per shift. Whether that was just the desire to be the best, or whether bonuses or bets were involved, I do not know. The inspector's shifts started one to two hours earlier than the contractor's crews, so inspectors would not be leaving just when the contractor's crew was trying to get an okay on the work performed that shift. Typically, the contractor's crews did not change shifts and some of the supervisors, foremen and workmen worked years on graveyard or swing shift, while others never worked those shifts. The Bureau crews of inspectors rotated shifts each two weeks or each month so that there was less opportunity for "buddy" relationships between the two organizations. The relationships were kept at arm's length.

I recall a few instances which might be useful to the reader in getting a good picture in his mind of the inspector's environment. Before sandblasting techniques had been improved, the removal of the laitance and calcium carbonate from the surfaces of the concrete was an arduous, labor-intensive task and the longer the concrete had aged, the more difficult the task. Too, the contractor was working over the full expanse of the foundation, so it took much longer to cover the entire area than it did later when the downstream blocks had been finished. So, there was a constant pressure from some of the cleanup crews to do "no more than necessary" and even to follow the inspector as he marked out the unsatisfactory areas with chalk, removing the chalk marks, but not the objectionable materials. In 1936 a show down came when I refused to okay a block that was cleaned to substandard and Ermie Stokes, the contractor's concrete superintendent called Bert Hall to come down to the block to review the requirements. Bert Hall and Ermie came into the block together and Bert took a quick glance at the work. "Ermie, you surely don't want to place good concrete on that, do you?"

Ermie grabbed a small wire brush from one of the crew and when he saw that the deposit was easily removed, told the foreman, in his own inimitable tone and emphasis, "You surely have to do better than that!" Cleanup was one of the items MWAK sought additional compensation for before the U.S. Court of Claims. The case was dismissed, however.

On another occasion, a rather newly-employed inspector was on a pour and as it neared 90% complete, the foreman left the block, telling the crew, "Just keep pouring concrete till I get back." We had an agreement with the contractor that only foremen or other supervisors could be given instructions about the work, so the inspector was unable to stop the concrete, even after normal height was reached. Many yards of concrete were added, heaping it instead of "dishing" it. We learned later that the contractor did not have another block ready to receive that concrete, but chose not to stop placing it anyway. I do not know at what level in the organization that decision was made, but I seriously doubt that the foreman had taken it upon himself to keep pouring. At that time Don Walter had been transferred to Friant Dam, so I was then in the role of liaison with the contractor's okay-chief. After learning the facts of the case, I went down to the block in question, marked the excess concrete, and instructed that it be removed at the contractor's expense. The next day Joe Harmon, the contractor's okay-chief came to me with a confession and an appeal. "Yes, you're right," he said. "We shouldn't have done it. But if you let us get away with it this time, I guarantee it will never happen again."

"Joe, I do not need your guarantee. When the concrete is removed I *know* it will never happen again!" And it didn't!

On another occasion one of the contractor's drains which were formed in the concrete to carry away the wastes from the cleanup of the concrete became plugged and all efforts to get it open and functioning again failed. The contractor's assistant superintendent Bob Niemen came to me seeking concurrence in abandoning the effort and proceeding with the construction. He told of the great efforts which had been made to get the drain open. "We have been working day and night, both from the inlet and from the downstream face of the dam to restore it, but it is impossible," he explained.

"I think you've been misinformed, Bob," I replied. I asked him to come out to the porch of the field office which overlooked the dam and directed his attention to the downstream face of the block in question. "Have you been using divers?" Bob blushed. The location was submerged by the higher stage of the river. The work was resumed.

After the War, Kaiser interests built automobiles at Willow Run,

Michigan and Steve Girard was the general manager of the entire activity. Later he became chairman of the board of the Kaiser Co. When he came to the dam to visit, he commented to Mr. and Mrs. Banks, his wife's parents, that quality control was the most demanding element of his responsibility at Kaiser. He added, "Vaughn was right!" so Mrs. Banks reported to me.

But for the most part, the job ran smoothly and everyone cooperated quite well. The problems encountered were perhaps to be expected, considering the inherent nature of mankind, the scope of the undertaking and the pressure to get it completed correctly, on time, and at a profit!

CONTRACTORS' ORGANIZATIONS

The MWAK partners took a very active interest in the dam activities and so resided in Mason City, some for the duration of the contract. Silas Mason, Tom Walsh Sr., Guy F. Atkinson and E.L. and

W.E. Kier were there and down on the job frequently enough for us to know who they were. I was not personally acquainted with any of them, however. Silas Mason did not live to see the work completed—he died of a heart attack in the hospital at Mason City on April 14, 1936.

At the outset, H.L. Myer was the general manager and Harvey Slocum was the general superintendent. They saw eye to eye all right, but they seemed just waiting for the other to blink. Not exactly a harmonious relationship. Harvey often talked about the necessity to go and "talk to god" about problems on the job. It was some time before I realized he was referring to Mr. Myer. About the time Mr. Mason died there were a number of changes in the contractor's management. Myer resigned, Slocum went off to India and the Bakra Dam, and George H. Atkinson then was general manager. Francis Donaldson was the chief engineer for MWAK until early in 1936. He was succeeded by C.D. Riddle, whom I knew. Robert L. Telford was the cofferdam engineer and superintendent of the river diversion activity.

Aaron Burrows was the cofferdam superintendent for part, if not all of that work. And there were others of "middle management." Frank "Pop" Smith was excavation superintendent and could be

Jim Butler standing on his head, smoking a cigarette atop the suspension bridge tower. I wonder, are his fingerprints embossed in those grips? October 10, 1935.

Edgar Kaiser, General Manager for CBI, ca 1960.

found down on the job at any hour of the day or night in the years of 1935-1937.

One evening Jack Walsh was trying to get one of the trestle footings on the bedrock cleaned up for concrete and was having difficulty because of a stream of dirty water coming over the cliff. When Frank Smith came by he sized up the situation and told Jack, "Go up there and tell whoever is causing that flow of water to stop it!" Jack was halfway up the cliff when he yelled, "And tell him Frank Smith said so!"

Jack came back a little while later, a displeased look on his face. "What did he say?" asked Frank.

"Well, he told me what to do, and where to go. And he don't know who Frank Smith is!"

I think perhaps he found out!

Jim Butler was the rigging boss—a real high climber—colorful, too. He once stood on his head on top of the center tower of the conveyor bridge just after its completion.

Waller was a garrulous pile-driving boss on the riverfront—another colorful character. Referring to somewhat inept pile bucks, he remarked, "They aren't very good with a pike pole, but you should sure see them spear herring!"

Bob Atwood was also on the riverfront, involved with barges, trestles, etc., although I don't recall his exact position. K.L. Parker was in charge of grouting and he, too didn't restrict his time on the job to the day shift. Johnnie Tacke was the concrete "boss" and Ralph Hawkins, Elmer Johnson, Max Pierce, Mel Russell, "Blackie" Coffer and Park Savage were part of Johnnie's shift bosses or "walkers" as they were called. Ermie Stokes was a carpenter "boss" for a while—there were others, too. My contacts with the work did not include the aggregate and concrete producton, the railroad operation, the cement handling, the trucks, the shops, the services, the camp—all of which were managed throughout the contract life by people not known to me. When CBI was the successful bidder, a new management team came to the dam and most of the MWAK top team were gone from the scene. But since CBI was part MWAK and part Six Companies, many of the residents in Mason City stayed on, awaiting their chances for employment with them. They wondered whether they would have to vacate the home in which they lived to make way for others who would displace them. Some stayed, of course,

but there were many changes in personnel. CBI brought a new management team to the dam with Edgar Kaiser the general manager, Clay P. Bedford the general superintendent and Bob Niemen the assistant superintendent. I am not sure how the lines of authority ran, but the relationships on the work seemed about as follows. Russ Hoffman was the superintendent of the dam and pumping plant and G.L. Dutcher may have also served in that capacity part of the time. Walking bosses—"walkers"—for the concrete work included Johnnie Tacke, Mel Russell, Malin, Max Pierce, Park Savage, Blackie Coffer, Elmer Johnson, Bob and Ernie Meyers, and Finley. Bob Switzer and Carl Watt were carpenter superintendents on the job and in the shop or panel yard respectively. Ozzie Mickelson was in charge of all pumps, Claude Bacon the gates, Armstrong and Scotty Wright the ironwork, Greenwood the grout, Johnnie Hallet the reinforcing steel fabrication, "Popeye" Baker the penstock trashracks, Roger Greenburg and later Steve Girard were in charge of all pipe installations, with Chris Ottmar and Larkin assisting. There were many changes in personnel on the job and my mention here of the above-named men is only a smattering of the organization that made the job successful.

When the right power plant was being constructed under the "cost plus" work orders, I had almost daily contact with Mr. Ray Dycus, who was then the office manager in the CBI office. I verified, reviewed and approved orders, bills, receipts and payrolls for payment by the Bureau.

The desk I used there for convenience was in the large room where the contractor's records and files were located. The work on the dam was nearly finished and the contractor was disposing of plant, equipment and supplies—records, too. In that connection I must call attention to one of the most remarkable exercises in management I have ever seen. The CBI controller, Mr. Joe Reis, from the Oakland, California office of the Kaiser Co. came into the records room with Mr. Dycus and inspected the materials in the many file cabinets. There must have been over 250 full drawers. They were working just adjacent to my desk, so I could not avoid overhearing their discussion. In not more than two hours, as they opened every one of those file cabinets and read off the code numbers on the folders, they decided the fate of every record of the CBI activity stored there at the dam! Decisions centered on what would be destroyed—a large list, indeed—and what would be stored at the large Odair warehouse near Coulee City, and what would be shipped to the Oakland office. I was aware of the CBI cost system from my contact with the Extra Work activity, but I certainly gained a higher regard for the system after witnessing that encounter.

Now, scanning my memory of those 1934 to 1942 building years, the strength of the contractor's organizations, both MWAK and CBI is very prominent. They had the financial, physical and moral strength to get the job done. But the management and control of those myriad activities—for those times—was something that they should have been proud of. Perhaps all of the partners in those companies have departed, but the legacy they established in the "Can Do Department" remains.

I marvel at the ability to get what was needed—now! Skills, machinery, equipment, supplies, teamwork, manpower—all seemed to be at their fingertips. They kept the wheels turning with precision against all adversity. If something went wrong, they fixed it and went on. *Remarkable!* Someone should tell that story, I think.

OFF THE JOB RELATIONSHIPS

Looking back over the events of those years at the dam I have the feeling that the social relationships between the contractors' people and the Bureau, while harmonious, was one of reserve. There may have been exceptions of course, but they do not loom up in my recall of the scenes. Perhaps the ground rules might have been set early in the project. After the concrete placing for the bridge piers had started, the contractor found that the Bureau really intended to have the quality of the work up to the standards prescribed in the specifications—perhaps a situation it had never encountered before.

A letter from the contractor suggested that perhaps the inspection requirements could be less onerous if the inspectors were supplied with occasional bottles of whiskey and entertained now and then as guests at social gatherings hosted by the contractor. The letter outlining the suggestion from brother John of the Seattle office was addressed to brother George who was running the job out on the riverbank, but in error, (?) the letter was dispatched in an envelope and for some reason, addressed to the *Bureau* in Almira. After reading the contents, Mr. Darland handed the letter to George. His face reddened as he, too read the letter. He was astonished! "Vell, I just don't know vat Yon vas tinking about!"

Mr. Darland, however, briefed George on what *he* was thinking about!

We inspectors may have been topics of lengthy discussions by the contractors. I recall one exchange in particular. I had been absent from the bridge pier work for a few weeks but was to be on a shift for a three-shift pour. I saw George in the mess hall and remarked to him, "Well, Mr. Johnson, I will be back on the job tonight."

"Dat's fine, Mr. Downs," he replied. "We'd radder deal wit debils dat we know dan wit debils dat we don't know!"

At Christmastime in 1935 Harvey Slocum brought a box of gifts over to the field office and left them. "Here's a few goodies for the inspectors. *Some of them* have been very helpful!" There were cigarettes and candy.

When the consulting board was on the site in 1934 or 1935, Mr. and Mrs. Mason invited the board, Mr. Banks, Mr. Miner and Mr. Darland over in the afternoon for tea. Mr. Banks related later that some of the guests were not accustomed to bourbon "tea." It was at that tea, incidently, that Mrs. Mason told Mr. Banks that she had named one of her Kentucky thoroughbreds for him. She named it *He Did*.

"I'm sorry, I don't quite understand," said Mr. Banks.

"When you finish the dam, you will!" she explained.

At small staff meetings Mr. Banks mentioned some of the happenings at social events hosted by members of the contracting family to let us know the tenor of the contractor attitude toward the relationships between the organizations. I think however, that he always held a very proper view of his relationships with contractor personalities. Mrs. Banks told Margaret and me of a time after Mr. Banks had retired. They were visiting in California and were house guests of their good friends, Mr. and Mrs. Guy F. Atkinson when Mr. Banks became ill. "Get me to a hospital," he pleaded with her. "I don't want to die in a contractor's home."

By the time the Kaiser Co. people arrived on the scene, the practices and standards of construction acceptable on the dam were well known and there were few show downs. The contractor was not above subtleties however, to let us know that some of our interpretations of the "specs" were slightly onerous. On one occasion a delegation delivered Mr. Banks two drawings showing how the contractor planned to perform certain work. It was all in jest, and Mr. Banks went along with the gag. The drawings showed the most complicated procedure

Mr. Banks reviews a CBI drawing with Mr. Darland, Mr. Miner and Mr. Hall. August 7, 1940.

Mr. Niemen, Mr. Reis (or Miller?), and Jack Walsh point out that the drawing was a hoax. But it made a point! August 7, 1940.

and refinement imaginable. Mr. Banks called the photographer to stand by for a picture, and escorted Mr. Miner and Mr. Darland down to the job where Bert Hall joined them to review the contractor's plan of action. When Bob Nieman, Jack Walsh and Joe Reis (or was it Mike Miller?) showed up, the jest was acknowledged—but the point was made. The drawings were numbeed 4QU and 4QU2, I remember.

WORKING CONDITIONS

By the time the MWAK camp with bunk houses for the singles and a great mess hall for any desiring to eat there neared completion in the fall of 1934, employment rolls began to expand. The workers, working stiffs as they were called, came to the job site any way they could get there. And since the Depression of the Thirties was still a scourge among us, many had been ill housed, ill clothed and ill fed.

As to the latter, the MWAK mess hall was the answer! The food was of excellent quality, prepared by fine chefs and served family style at tables for eight. Food was brought to the table by clean, attentive attendants called ''flunkies.'' The utensils, dishes, plates, tables and floors were clean and sanitary. Harvey Slocum, the MWAK superintendent didn't think he could get a full day's work from an underfed man and no one who ate at his mess hall could ever say he hadn't had enough to eat. The men worked hard, but they were well fed and the food was *good!* Slocum had many attributes. He knew good food and he could prepare it himself. Once when he and his guests were dining at the Davenport Hotel in Spokane, he called the waiter to his table and complained, ''These biscuits are not fit to eat!''

The waiter relayed the message to the chef, who sent Slocum another serving of biscuits and the reply, ''If you think you can make better biscuits, just come on back!'' Excusing himself to his guests, Harvey walked back to the kitchen and donned an apron. He reappeared at the table some time later with a plate of really good, hot biscuits!

I ate at the messhall regularly for two months, and later took my family and guests there on occasion, as did many of the men. It was an experience and a treat. That mess hall served the most and the best for six years. Alas, it is no more.

Things were not so comfortable down on the job. It was either hot and dusty, or cold and wet, or freezing. The work went on 24 hours a day, with three shifts of 8 hours each; day, swing and graveyard. There were no coffee breaks, but at mid-shift time was taken to quickly consume the lunch brought from home or the mess hall, carried in a tin bucket or brown bag. Water barrels, canteens and evaporative water bags filled with drinking water satisfied their thirst. Sanitary facilities were primitive and when the work force grew in numbers, toilets were located convenient to the activity. Chemical toilets had not yet come on the scene, and the Columbia River passed by so conveniently.

When the workmen were first employed, they were issued a brass badge about two inches in diameter, individually numbered with MWAK and the employee number deeply incised so it could be clearly read from a distance. Incidently, at the 50th Anniversary gathering at the dam on July 16, 1983, I saw the polished "MWAK 2" proudly worn on the lapel of Mr. Guy F. Atkinson's grandson, Ray N. Atkinson and "MWAK 2838" worn with equal pride by a man who had been a carpenter while the dam was being built. That badge was your identity when you checked in the bunk house, the mess hall, the entrance gate, the payroll office, the time shack and with your supervisor. You didn't leave home without it! Indeed, you couldn't get into the work area unless you had your badge in view. We Bureau employees also had numbered badges—small, silvered shields. Mine was #10. These were later replaced with small identification cards sealed in clear plastic, printed with your employee number and your photograph, which gained you access past the guard at the dam entrances.

At some stage in the job, workmen were furnished hard hats for protection from falling or flying rock, tools, etc. The first hats were of laminated plastic/cloth material with a headband and harness to take up the shock of a direct hit. This made you a member of the Turtle Club and not a statistic. Later, aluminum construction replaced the plastic hats. The contractor also furnished heavy, rubber, knee-high, open-topped boots to workers in the concrete. The Bureau furnished the inspectors with knee-high rubber boots which laced up. Slickers too, were available for the wet work or for the days when it rained hard. When it was cold, warmth was nearby in "salamanders," open steel barrels with ventilation holes cut around the lower sides where fires were kept burning, using scrap lumber. There, the foreman could watch the progress of the work and the drillers and powdermen—powder monkeys—from the rock excavation crews and the carpenters,

pipefitters and welders could warm their hands when they got stiff with the cold.

In order to avoid acute storage problems for forms and for work efficiency, the contractors took advantage of the permissives of the specifications so that the maximum differential in height between any two adjacent blocks or panels of the dam would not exceed 15 feet normal to the axis of the dam nor more than 30 feet parallel to the axis. Exceptions were made for the diversion slots (MWAK) and at the ends of the spillway. Thus, typically the forms were raised and placed, ready for the next lift before the preparations for concrete in the adjacent block would begin. This stepped construction required much "climbing of the walls" and heavy ladders were built to accommodate and they were of the lengths needed—up to 30 feet for moveable ones. Through traffic for moving from one area to the next across the dam was accomplished via catwalks that spanned the gap between high blocks. The carpenters, pipefitters, welders, ironworkers, concrete, cleanup and placing crews and vibrator operators all carried their tools on their shoulders—climbing the ladders, rushing along the narrow tops of forms, or across the catwalks from one location to the next. And the engineers and inspectors as well as the shift bosses for the contractors did the same or viewed progress from the deck of the trestles. A few—Ermie Stokes, the wild, loud carpenter boss, for example—rode a skip from and to the deck of the trestle when he couldn't be heard as he shouted above the whistles, the high-pitched scream of the little, air-powered pumps in the sumps, the rumble of the concrete trains, the noise of the jackhammers, trucks and men.

Ermie Stokes was an ardent fisherman and Owhi Lake brook trout were his favorites. So it mattered not that the lake was on the reservation and was closed to fishing by non-tribal members. Ermie fished at night. Many a night, though, did some of the fellows from the dam go up there and chase him along the shore. They never caught him however, as he fled from his pursuers. The next morning down on the job, he would tell how he could "outrun those _____ Indian police!"

Stairs were located at a few points and the ladders were not caged until much later, nor were there any landings where one could pause to rest. I recall my fright when, winded in my ascent up a long ladder, 50 feet above the rocks below, I stopped to catch my breath. I looked up to see a carpenter, his big box of tools slung over his shoulder, starting

Drillers and others on steep slopes used ropes, "hand lines," for safety in moving from one level to another. August 1935.

down from the top! There was barely enough room for him to pass me as I clung to the ladder with tooth and toenail!

The drillers and steel handlers—nippers—also had uneven terrain to traverse on the abutments and in the deep, weathered zones of the foundation. Where the rock was too steep to pass unassisted, ropes were fixed to anchors above and such were used to climb from one level to another. Dynamite was delivered by truck each shift, shipped separately from the blasting caps as a precaution. The powdermen would then punch a hole in the stick of dynamite, insert a blasting cap and wrap the wires around the stick so they wouldn't be dislodged, (electric blasting caps). Moving over to the just-drilled holes, they air-jetted all loose dust out of the holes and loaded them with the dynamite and primers. They next tamped dirt or rock dust into the hole and connected the wires to the blasting circuit and headed for the whistle to put all on notice that, just before shift change, there was going to be a blast. The area would clear of workmen, the big shovels and all equipment was moved back to what they considered a safe distance, and then the blast. The foreman and the powdermen would listen carefully to see that no round of charges failed to fire! Millisecond delays for blasting caps had not come to the dam yet, and the blasting

caps' delay time was varied—the longer the cap, the longer the fuse within and the longer time in fractions of seconds between the throw of the switch and the explosion.

Grant Wood, the famous artist who painted "American Gothic" never came to paint the "Dam Scene." Had he done so, he would have lined up the subjects in a long row in the foreground. Instead of wan faces and bareheaded, holding pitchforks, he would have painted some in hard hats, a surveyor with a transit, a batch plant operator with his scales, a laborer with a muck stick (long-handled shovel), a steel nipper with an assortment of sharpened drill steel, a driller with a jack hammer, a carpenter with a hand saw and hammer (no portable power tools yet), the cleanup hand with a broom, a bucket of sand and an air-water jet, the concrete hand with a vibrator, a shovel operator (just a mucker with a bigger shovel), the pipefitter with a wrench, the welder with his acetylene tanks and torch, the riveter with tongs and red-hot rivet, the cement finisher with a trowel, and an inspector with specifications in hand—almost a score in the painting, but there were many more on the job! They were the ones who *made* the dam. And in the background Wood would paint the mighty river to be tamed, the distant steel and cement mills, the railroads, the trucks upon the highways, the designer

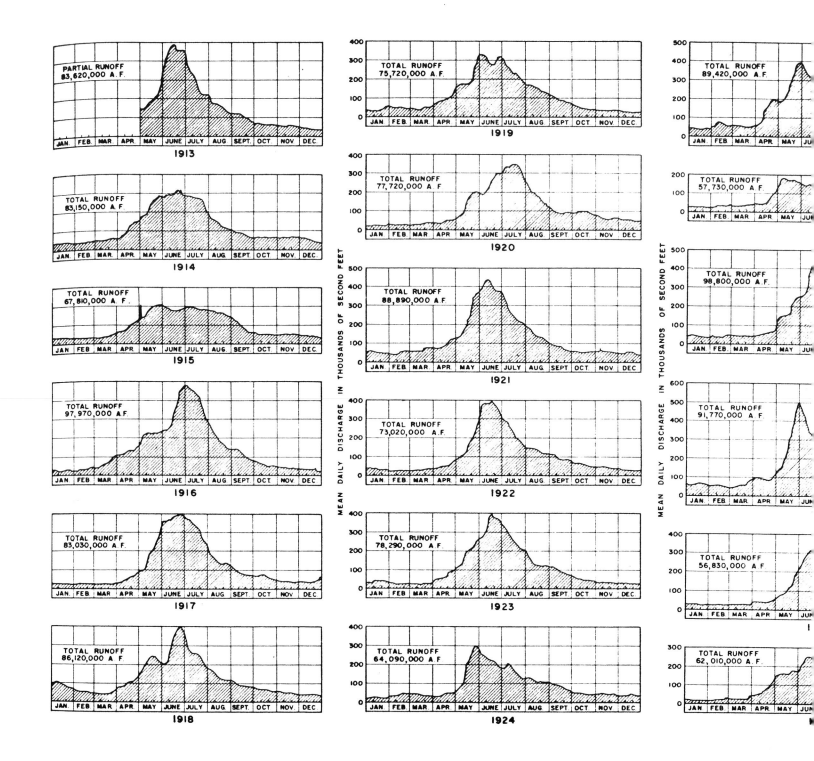

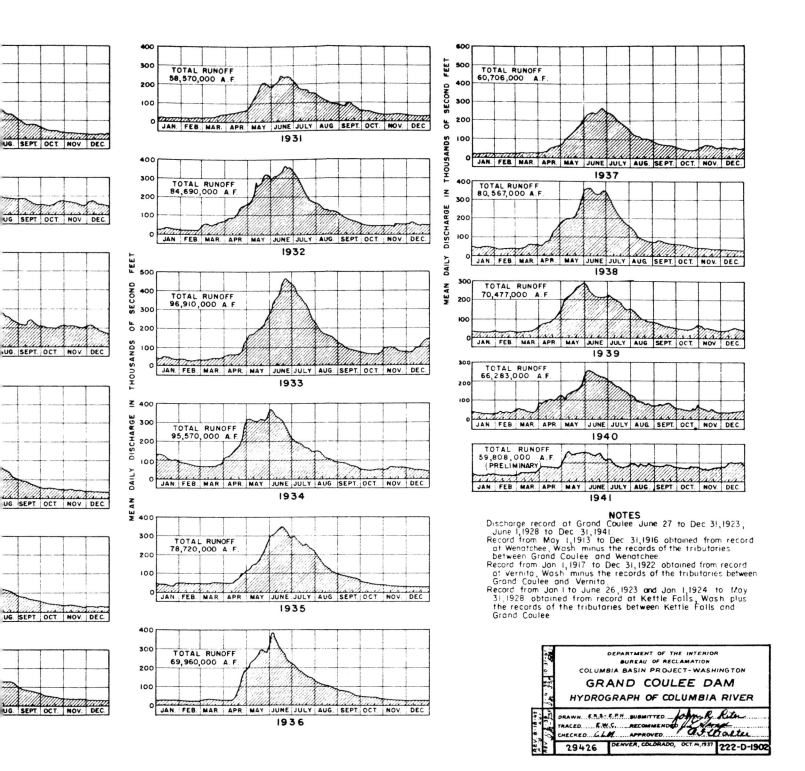

NOTES

Discharge record at Grand Coulee June 27 to Dec. 31, 1923, June 1, 1928 to Dec. 31, 1941.

Record from May 1, 1913 to Dec. 31, 1916 obtained from record at Wenatchee, Wash. minus the records of the tributaries between Grand Coulee and Wenatchee.

Record from Jan. 1, 1917 to Dec. 31, 1922 obtained from record at Vernita, Wash. minus the records of the tributaries between Grand Coulee and Vernita.

Record from Jan. 1 to June 26, 1923 and Jan. 1, 1924 to May 31, 1928 obtained from record at Kettle Falls, Wash. plus the records of the tributaries between Kettle Falls and Grand Coulee

DEPARTMENT OF THE INTERIOR
BUREAU OF RECLAMATION
COLUMBIA BASIN PROJECT-WASHINGTON

GRAND COULEE DAM
HYDROGRAPH OF COLUMBIA RIVER

DRAWN. E.R.S.-E.P.H.   SUBMITTED. John R. Riter
TRACED. E.W.C.   RECOMMENDED
CHECKED. G.L.M.   APPROVED. A.F. Walter
29426   DENVER, COLORADO, OCT. 14, 1937   222-D-1902

73

with slide rule and drawing, the contractors around the table sharpening their pencils, and the superintendents and shifters shouting and waving their arms, and above, the vision of the dam complete with that beautiful, white water over the spillway.

SAFETY

Accident prevention was becoming a greater concern in industry generally, and the specifications for the dam required the contractors to, "At all times exercise reasonable precautions for the safety of employees on the work," and to report all lost time accidents promptly, monthly. But it was not until August of 1937 that the Bureau initiated an accident prevention program of its own. It was appropriately called the Safety Program and on the dam, Charles Parsons was designated the Bureau safety engineer for the dam. The chief enginner also designated a safety engineer. The contractors established a Safety Department and high on the abutment the mammoth sign broadcasted for all to see and take to heart, "SAFETY PAYS."

The Bureau generally concentrated its efforts to promote safer jobs by urging and cooperating with the contractors in establishing safe methods and practices. Accidents that did happen—not few in number—were investigated as to cause and the cases were used in review in safety meetings with foremen to prevent repetition. Accident report forms were standardized and safety was no longer considered an afterthought. Nevertheless, numerous accidents did occur, and fatalities during the construction of the dam totalled 79, if my recall is accurate. Injuries and deaths resulted from falls, falling material, failure of scaffolding and form ties, blasting accidents, crushing under or alongside moving loads, drownings, equipment failure and vehicular accidents, etc. On one occasion when the Bureau safety engineer came to the dam to promote safety on the job, it seemed that he was serving a lost cause. To his horror and that of the rest of us building the dam, that was the worst week of all, safetywise. By the end of the week six men had died in *preventable* accidents! Many of the workmen had very little experience when they came on the job, and some of them neither had the skills nor the agility and balance—of tight rope walker's variety—that would have been so helpful in their

roles as constructors. So many did not stay, "Not my bill of goods!" But others stayed and tried to stick with it, to their sorrow. Rumors were recurrent over those years of "Men buried in the concrete." I swear that they were all false! Perhaps they were started when a workman (purposefully?) lost a rubber glove in the concrete adjacent a formed surface. The form was removed and what appeared to be a hand rising from the concrete startled onlookers. But that's all it was, just a glove!

Stairs at left for access to or from block 40 where it was joined to the cofferdam by concrete. Note also the vertical ladder up the east face of the block. Lower 5' by 7' inspection gallery at elevation 900. At elevation 950 above is the 5' by 11' 6" gate service gallery. Large, circular tunnel is for control cables between right and left powerhouses. May 24, 1937.

CHAPTER IV
THE COLUMBIA RIVER

THE RIVER RESISTS

The builders of the Grand Coulee Dam had a major obstacle to overcome that the builders of the pyramid of Cheops did not. Here was a major river with the bedrock covered with a great volume of earth and rock. But the Columbia River had predictable flow characteristics as shown on the hydrograph on pages 72 and 73. And the clay overburden lying upon the bedrock had been compressed by the overriding continental ice of the last ice age and included very large boulders as well as lesser river or glacial-worn rock, making the task ahead much more hazardous. And preliminary excavation disclosed through the slides encountered that the clay had very little resistance once disturbed.

I look back with great respect and admiration to the contractors who staked their personal fortunes on their confidence that the river could be tamed. That took courage!

The engineers charged with the responsibility of the diversion scheme and the cofferdams had an unprecedented task of major proportions. The task—to design a cofferdam that would do the job, that could be built in a short time, that would hold out the river effectively without excessive seepage, that in part could be maintained through three major floods, and that wouldn't cost too much, since it was to be only temporary and would have to be removed. Lastly, the cofferdam designers had to rely on the capabilities of the cofferdam constructors on the job to get it built as designed. That responsibility would weigh heavily. It is no wonder then that Bob Telford was called "The Cofferdam Engineer," as he was down there on the job so much of the time.

Fortunately, the cofferdam included a substantial factor of safety or more luck than could be expected because it proved impossible to drive the sheet piling to bedrock as planned. Bedrock was at elevation 880, some 55 feet below the low water level where driving started. But the piles came to a halt at about 30 feet of penetration and many encountered boulders and stopped short or split out at lesser depth.

The failures of elements of the cofferdams from pulling out of the interlocks from internal load of the backfill, while of concern and added cost, did reassure the principals—or should have—that the cofferdam was not over-designed. Also, the contractors were blessed with no abnormally high discharge peak or volume of the river flow while the cofferdams and river "racheting" activities were underway. Peak discharge and volume for those years were as follows:

	max. cfs.	volume ac. ft.
1935	354,600	81,400,000
1936	387,000	70,800,000
1937	271,000	57,500,000
1938	364,000	81,960,000
1939	297,000	69,280,000
1940	265,600	66,490,000
1941	170,000	53,700,000

The average annual volume of record has been 79,620,000 acre feet. The highest volume of record in any year was 103,800,000 acre ft. The lowest volume of record in any year was 53,700,000 acre ft.

From winter low water to flood stage the river rose 39 ft. in 1935, 32 ft. in 1936, 31 ft. in 1937, but in 1948 rose 61 ft.

RIVER CROSSINGS

Quick access across the Columbia River was essential for the orderly progress of the work and of course an early completion of a permanent bridge was first on the schedule. The bridge was to be designed by the State Highway Department. It planned a beautiful suspension structure located at the foot of the steep grade down the hill to the Government Camp. When Mr. Savage reviewed those plans he urged that the bridge be moved 1,000 feet farther from the spillway and to use a cantilever design. Fortunately so, for the sliding ground on the east bank of the river and the future construction of the Third Power Plant would have destroyed the bridge.

The State Highway Department made the designs and bids were invited first for the main piers and later for the bridge structure, abutment piers and approaches. The low bidder for the piers was not experienced in work of this type, but had been doing general building construction where foundation problems were not complicated. But

Highway bridge in foreground open to traffic, suspension bridge next upstream for foot traffic, then barely visible, the cable for the Seaton ferry shown at the east shore. The pile bent bridge crossing to the west cofferdam supported conveyor belt for transport of excavated materials.

The suspension bridge crossing the entire site carried a conveyor belt for aggregate transport to the west mixing plant and a steel pipeline for transport of cement to the east mixing plant. January 1936.

the company wanted the work, and furnished bond, and so was awarded the job—to its later regret, I assume. A cableway was constructed with the headtower on the right bank and derricks moved in to serve the two piers, but high water came before it could be tolerated and the job got off to a slow start. After it receded, the work of constructing and sinking the piers to bedrock began. Pneumatic caissons were a new experience for the contractor. Finally, after reluctantly employing caisson supervisors such as Miles Philbin and sandhogs, they got the piers to bedrock. Thereafter the piers came along at a good pace, but the job was late and the contractor, J.H. Pomeroy and Co. had a delayed start erecting the bridge superstructure. The bridge proceeded without any hitches, until all was in readiness to close the central span onto the right bank cantilever section when it appeared to be about nine inches too long. Well now!

But it was worse than first surmised! The sliding of material along the right bank—which had caused abandonment of the planned railroad extension, had caused the pier to tip toward the river, requiring immediate action. The bridge contractor was directed to construct an open steel caisson about the distressed pier, removing the earth therefrom, after having anchored the span against further

outward movement by heavy cable stay lines farther shoreward and by struts across the opening in the superstructure. With the burden relieved, the pier eased back into position, concrete encasement of the pier followed, the span was closed and the bridge opened to traffic—but it was late.

Early on, the contractor for the dam, MWAK also constructed a suspension bridge across the river, somewhat upstream from the proposed highway bridge. This permitted workmen and others to cross without interference with the contractor's operations and without resort to Seaton's Ferry or the bridge pier contractor's cableway. "Cap" Tuttle obtained a permit to operate a motor-operated ferry on May 26, 1935, and Seaton's Ferry was discontinued. Also, MWAK constructed a heavy-duty pile-bent timber bridge, but the deep, fast water and the ice in the river in the winter put it out of service and it was removed. As the need developed, MWAK added barges and boats to respond to the riverfront needs. Another pile-bent bridge was constructed to carry a conveyor across the river to the west cofferdam area for disposal of excavated material and for delivery of selected coarse material for cofferdam filling. That bridge remained in use for only a few months but transported over 2,000,000 cubic yards of material. It too, lost two of its bents to the flowing water but was repaired. The U.S. construction railroad was extended down the steep grade from the axis of the dam to the Left Powerhouse. The grade and lay of the land necessitated the use of a switchback and "Shay" locomotives. MWAK then extended the trackage on north, skirting the town of Coulee Dam by means of a tunnel through the granite mass at Fiddle Creek. A timber truss bridge supported by steel piling and protected by steel cofferdam cell enclosures supported the bridge. The sliding ground and the instability of the riverbank on the east side of the river dissuaded MWAK from building the railroad beyond the east end of the bridge—but it was used for truck traffic throughout the dam building—January 1935 to 1941.

DIVERSION AND CARE OF THE RIVER — Cofferdams

Diversion of the river around the job site by tunneling was not a viable option. The volume of the stream at time of flood was much too large and the runoff of summer snow-melt was retarded by the high mountains and in the many large lakes in the headwaters, so the flood runoff continued for three or four months each year. So the contractor planned to protect the work areas from the river by the use of a cofferdam system as sketched on this drawing on page 78. The numerous photographs show the contractor's progress throughout the life of the dam building. Readers referring to the dates on the photos may appreciate the pace of the job. Too, the size of the work may be visualized when the quantities of material used in the cofferdams by MWAK are noted. The steel sheet piling used totalled 18,606 tons, the miscellaneous steel weighed 2,822 tons, 18,500,000 board feet of lumber was required, filling of the cells took 1,502,735 cubic yards of select material and an additional 1,328,735 cubic yards of excavation was required The reported cost was $5,864,177.

The left bank cofferdam would encroach into the river only slightly at low water but would extend to sufficient height to keep the flood flow in 1935 and 1936 out of the protected west bank works where excavation, foundation preparation, and concrete placement would proceed. While so protected, the concrete would be built up above normal river levels, leaving 11 intermediate blocks low for the passage of river flows during the subsequent diversion of the river from its channel in the fall of 1936 and the 1937 flood season when the main

channel of the river would be closed off by cofferdams across the channel upstream and downstream of the dam. Sufficient room between the cofferdams would permit the excavation of riverbed material from the dam foundation, hopefully without incident. Construction within that area would then bring the concrete blocks to their prescribed height under the change order and the cross river cofferdams would be removed. Following that, the planned subsequential closing of the "slots" in the west part of the dam would complete the MWAK contract. In order to advance the excavation from the right bank a small cofferdam faced with timber piling was planned to protect the area being excavated until the cofferdam was overtopped in 1936—elevation 965— with the rising floods. The entire cofferdam plan was successful—but not without incident. *And* it was a *major* undertaking.

The preliminary excavation of the overburden disclosed that beneath the mantle of earth, sand and gravel was a deep deposit of indurated glacial clay. The clay had low shear strength as disclosed by test and in observations during the excavation, and it contained discontinuous lenses of sand and gravel. Also the mass included scattered cobbles and boulders, some being very large. These characteristics of the material had an inherent hazard and difficulty for the cofferdamming scheme.

The west cofferdam consisted principally of steel sheetpiling cells joined in a continuous structure with the main section normal to the axis of the dam and with the up and downstream arms curving off into the bank above the high water mark. The enclosed area protected all areas required to be excavated for the structure and for the powerhouse tailrace. Just beyond the limits of the dam, spaced over 800 feet apart, larger clusters of steel sheetpiling cells were constructed, which acted as anchors for the cross river sections later to be constructed. At those clusters also, the cofferdam was later extended to Block 40 of the dam, making that portion of it to serve as part of the enclosure, keeping the diverted river from flooding back into the work area to the east. The steel sheet piling was of the interlocking type, 15 inches wide, weighing 38.8 pounds per foot and was received at the site via truck and trailer from the rail head in lengths up to 80 feet long. "T," "Y" and "X" fabricated sections were used in tying the cell structures together for continuosity at the joints. Wooden templates had been constructed (after removal of the coarse mantle of heavy gravels and cobbles on the

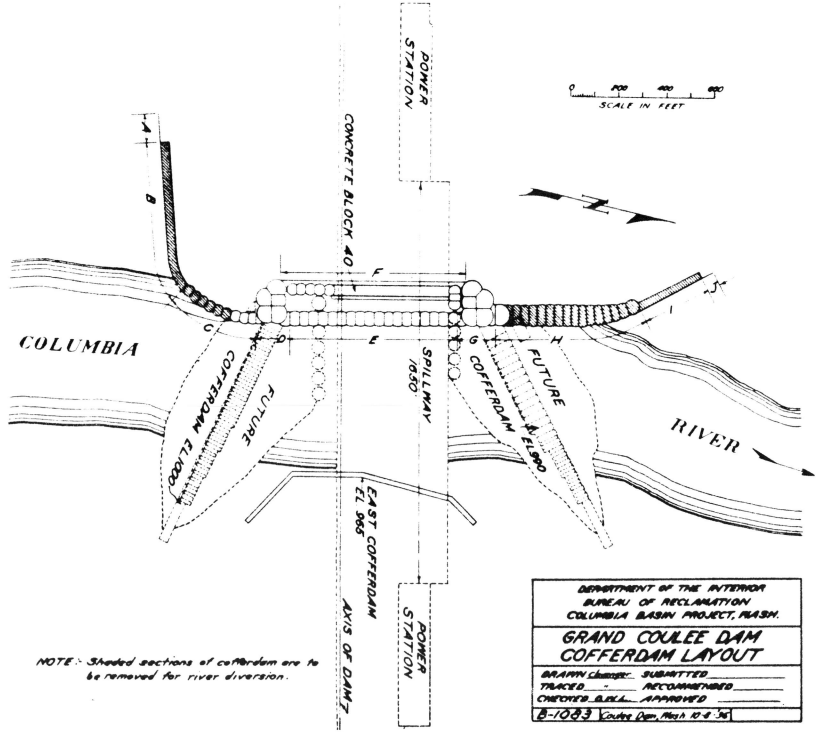

Grand Coulee Dam Cofferdam Plan. 1934.

riverbank) to guide the placement of the piling and to provide support as the cells took shape before start of driving. Succeeding sheet pilings were guided into the preceeding pile interlock by a "rigger" riding a sling or "bosun chair" suspended from the boom of a crane—the free piling being suspended from a second line from the crane. With a cell thus complete, the piles were driven into the overburden using steam-operated pile driving hammers. The first piling was driven on New Year's Day, 1935 if my memory serves me correctly. But driving the piling into the clay was difficult. The contractor had intended driving the pilings to bedrock (about elevation 880) starting at near river level of elevation 935 but the driving friction encountered exceeded expectations—it being almost impossible to get more than 30 feet of penetration. Also when large boulders were encountered, the piling stopped, turned away, splitting out of the interlocks or in some instances (as disclosed when the base of the cofferdam was later exposed by the excavation of the adjacent material) the piling turned back toward the surface as driving efforts had continued.

Attempts to drive pilings singly was to no avail but progress was gained when the driving progressed around the cell driving each a few feet then on to the next in succession. The slow driving necessitated bringing more hammers into use, but cranes to serve them were not in excess. One of the young foremen then suggested that after setting the cell pilings in place, the hammers be suspended from hoists on trolley rails carried on moveable wooden gantry frames. Thus, the driving got back on schedule. After being driven, the piling was trimmed if need be, and extended to elevation 990 on the river side and the cells were filled to that level with excavated material, chiefly sand and gravel. By that time, forecasts portended a higher than normal peak on the spring flood and the cofferdam was raised another five feet as an added precaution. A few cases of splitting out of interlocks and failure of locally fabricated riveted joints caused the contractor to add strengthening measures on the outer face of the structure. But it served the need through the floods of 1935 and 1936 (peak flow of 354,600 cfs. in 1935.)

As soon as possible after the 1935 flood, the contractor began excavating to bedrock, an open but laterally supported trench about 60 feet in width and extending the full width of the dam. This trench overlapped the dimensions of the 50-foot-wide block (#40) of the dam and was located adjacent the existing cofferdam. In this trench which

had to be supported continuously as it was dug and while construction therein advanced, the bedrock for the block was excavated and the concrete for that block of the dam constructed—the shoring of the walls being removed from the bottom up as the block grew in height. Sheet piling cells were extended from the upstream cofferdam cell cluster and terminated in a concrete connection to the upstream face of Block 40. Connection between the concrete on the downstream face of the block however, required the construction of timber cribbing in the bucket of the dam and up the face of the dam soon after each lift of concrete was placed. The back side of the existing cofferdam was then supported against the newly-placed concrete and against the cribwork being built on its downstream face and bucket. The existing downstream cofferdam cell structure was also extended and connected to the cribwork. Then the remainder of the cofferdam lying westerly of Block 40 could be removed permitting the adjacent excavation to be completed and the concrete brought to desired levels for the diversion of the river later in 1936. After concrete in Block 39 was constructed, a timber crib was also constructed in the bucket as was done in Block 40 providing a timber crib cofferdam 100 feet wide.

The shore arms of the crossriver cofferdams began to take shape as the river receded after the 1936 flood. And in the forebay diversion channel to the slots through the west section of the dam, the contractor assembled the timber cribs for the river sections of the cofferdams which were there constructed in the dry.

Meanwhile, with the westerly portion of the dam completed for the next stage, the upstream and downstream arms of the west cofferdam were removed, causing the cribs to float. When the gap was widened enough, the cribs were towed into position in the main river channel. First the downstream and thereafter the upstream units were placed. The downstream cribs were sunk into place by filling the ballast sections with excavated rock. Next stoplogs were dropped into the sluiceways through the cribs—they had been left open while the cribs were being placed. Then the cribs were filled with gravel and sand, effectively closing off the flow in the river channel. The remaining cribs for the upstream cofferdam were then sunk in place and both the cofferdams were faced with a line of sheetpiling driven to refusal—then extended above the 1937 flood levels. Those facings of steel were

[Continued on page 83]

Construction of the West cofferdam—driving timber piling for templates and trestles, removing mantle of cobbles and gravel from river channel. At far right the first cell structure piling erected but not fully driven. January 30, 1935.

West cofferdam sheet piling essentially complete and back fill of cells almost finished. March 29, 1935.

The West Cofferdam construction in "full swing." Timber gantry at mid-point of cofferdam used to support pile driving hammers. March 1935.

West cofferdam was raised an additional five feet when forecast was received of higher-than-expected flood. East excavated area flooded. Excavation continues—disposal by conveyor system in foreground. June 20, 1935.

MWAK General Manager H.L. Myer (on the right) with the Company engineering team. Jim Foster, Robert L. Telford, Silas H. Woodward (consultant), George H. Atkinson (succeeded Myer as General Manager), Francis Donaldson (Chief Engineer), Douglas Riddle (succeeded Donaldson as Chief Engineer). Photo courtesy Robert L. Telford. October (?) 1934.

Behind the main cofferdam the construction of block 40 was undertaken using timber trusses to support the sheet piling. As excavation advanced, additional trusses were added below. After concrete placing began, the trusses were removed from the bottom up after earth fill was placed between concrete and sheet piling to avoid cave-in. September 27, 1935.

The timber cofferdam on the left was built to exclude the rising river only until the peak of the flood approached. That permitted time for much excavation to be performed. Note high water mark on west shore cofferdam from the 1935 flood. Suspension bridge on right carried aggregate to the West Mix plant on far abutment. Pipeline to transport cement to East Mix plant not yet installed on the bridge. November 2, 1935.

Block 40 and the cribbing in its bucket section appear as excavation progresses. Fill material for the cofferdam cells and for back fill adjacent the concrete in block 40 was transported across river and along the cofferdam via conveyor belt system. Concrete was delivered on the transport cars via a rail track from the west mix plant along the excavated forebay slope, (not pictured). April 2, 1936.

By April 29, 1936 the cofferdam was well supported against the monolith of concrete and cribs of block 40. The flood waters inundated the pit on the east bank, flooding over the timber cofferdam as planned.

Block 40 and its cribwork rose out of the pit and the adjacent cofferdam was removed after the crossriver cofferdams were above the river level. The river now flowed through the west abutment portion of the dam. January 6, 1937.

Late in September the "slots" through the dam were "readying" to receive the diverted Columbia River. September 27, 1936.

locked into the cell clusters of the west cofferdam remnants, making a continuous temporary dam to protect the entire area east of block 40 from flooding during the 1937 flood. At least that was the plan.

Meanwhile the removal of the shore arms of the west cofferdam downstream from the major cell cluster continued. It is presumed that the pulling of those piling and the accompanying excavation of the riverbed at that point may have disturbed the clay underlying the main cell cluster. Or perhaps more likely, exposed a layer of sand within the clay. Soon the seepage through the cofferdams increased beyond the capacity of the dewatering pumps but more pumps were promptly installed. By March 17 the leak through the cell cluster began to alarm the contractor and efforts to stem the flow under the cofferdam went round the clock. Truckloads of rock, sand , gravel, clay, brush, hay and even a few old mattresses were dumped into the river adjacent the cell cluster to plug the flow but the inflow continued to gain on the pumps so more were added. And the dike constructed across the bedrock to the east of the trouble pooled the water to a depth of almost 20 feet while the excavation of rock and the concrete activity to the eastward continued. Then about 10 a.m. on the 18th of March the steel piling forming the southeast cell of the cluster split open, spilling the fill material into the excavated area, threatening the entire work in progress. The contractor, alert to the possibility of the complete failure of the cell cluster stopped all other activity and cleared the areas below the river level of all men and movable equipment. But the persistence of the effort paid off—the leakage was stopped and the cell cluster was sealed by injecting cement, bentonite and shavings into numerous holes drilled to the base of the cofferdam piling. Those efforts then returned the inflow through the cofferdams to normal—about 2000 gallons per minute as compared to an inflow of perhaps 29,000 gallons per minute at the maximum. To strengthen the cell cluster during the river rise several additional cells of steel piling filled with gravel were constructed adjacent the ruptured cell. The tower for the aggregate conveyor bridge was then supported with large concrete-filled pipes extending to bedrock.

The contractor reported the grouting operation alone consumed some 821 tons of cement, over 71 tons of bentonite and 12,000 cubic feet of sand, shavings and sawdust. But after a few frantic weeks of sleepless nights and the expenditure of some $400,000, the cofferdam—and the contractor—were saved.

But before those activities had concluded, one of the very high cofferdam cells that supported the tie to the upstream end of block 40 split open and spilled its contents also. But leakage did not seem to be

Cribs for the crossriver cofferdams had been built upstream of the dam ready to float out on the "tide." September 27, 1936.

The cross river cofferdams grow while the shore arms of the west cofferdam are removed, opening larger channels for the diverting river. December 2, 1936.

Soon after the crossriver cofferdams were sealed with the earthen and rock fills, the adjacent enclosed pool of water froze over. Work on the raising of the cofferdams to needed height continued unabated. January 6, 1937.

The current through the channel to the dam increases as the blockage of the main stream continues. November 20, 1936.

Removal of shore arms of the west cofferdam continued while the raising of the cross river structures proceeded. January 8, 1937.

The Columbia River was frozen over again after passing through the diversion channels through the dam. It froze over also in the winter of 1935/36. The blocks in the dam left low were to permit passing of the 1937 flood flows. This view shows how cofferdams were connected to the dam (blocks 39 and 40) to exclude the river from the work area. Mason City and the Brett gravel pit in the background. January 28, 1937.

a major cause of that rupture. There the repair consisted of a concrete retaining wall constructed parallel to the axis of the dam and rising to the westward as support for the endangered cells. Against that concrete wall, backfill of earth and rock supported the cofferdam elements. I don't know what recognition Frank Smith and Bob Telford received for their efforts—but I am sure they got a few grey hairs. That was an aging process, just to watch!

When the structure work within the cofferdammed areas had been completed and the work accepted, the areas within were flooded and the cofferdam removed to specified levels and disposed. The scrap steel was sold—a lot of it going to Japan. The removal of the timber cribs from Blocks 39 and 40 and the connecting cofferdam elements deposited material in the bucket which was removed carefully, using dragline buckets (shielded with timber) from the floating dredging fleet. While so doing, extensive deposits of debris were found and removed along the length of the bucket. This finding raised concerns as to the condition of the concrete of the bucket.

[Continued on page 88.]

The underflow through the cofferdam undermined it, causing one of the cells to collapse. The field shack caught fire when the stove tipped over. Minutes before the collapse the shack had been occupied. Water in the foreground was from the underflow through the structure. March 18, 1937.

By mid-April the cofferdam had been restored to permit excavation to continue. Washington Air National Guard photo. April 16, 1937.

Excavation, concrete placement and cofferdam strengthening continue. The cofferdam cells at both the upstream and downstream ends of block 40 were supported to prevent blow-in during the oncoming flood. April 17, 1937.

The rising river diverted through slots in the dam overtop the powerhouse walls subjecting the reinforcing steel to vibrations in the current. Many of the bars (1¼ inches square) were broken by stress fatigue and had to be later restored. May 11, 1937.

MWAK Company concentrates its activity to finish the high trestle promptly to permit serving the construction area with concrete from both mixing plants. Serious underflow beneath the cells of the cofferdam cluster supporting tower for aggregate suspension bridge incurred costly delay in area where bedrock shown exposed. Seepage water was captured in sumps and discharged over the cofferdam back into the river. June 23, 1937.

MWAK concentrates efforts in low area of foundation where rock is not yet covered. August 4, 1937.

All bedrock now covered with concrete and cross river cofferdams being removed as river recedes. September 1, 1937.

On August 9, 1986 Mr. Robert L. Telford returned to the dam to reminisce and to see the mighty structure "in the flesh." He had not returned previously—having departed this area in 1937 after the diversion of the river for MWAK had succeeded. Mr. Telford, now 87, is the chairman of the board of directors of the Mason and Hanger—Silas Mason Co., Inc. I met him there on the riverbank and in the Bureau facilities we reviewed some of the highlights of those eventful years in our early lives.

Items foremost in his recollection were the "big mud slide" and the ice dam and how Francis Donaldson, chief engineer for MWAK convinced the partners that the idea would succeed; the great conveyor system for removal of the earth and clay from the foundation to Rattlesnake Canyon, and the March 17, 1937 failure of one of the major cofferdam cells. As to the latter he told me of the occasion in mid-March 1937 when the cofferdam crisis was getting all of the attention and he and Frank Smith were there with the MWAK partners. Frank Smith had just been asked which shift he wanted to supervise to save the structure when a live, 14-inch fish came through

the cofferdam. Frank replied, "Neither one!"

But he took the day shift and Bob Telford took the night shift. He recalled, too, his response to the MWAK partners when in February 1935 the pile driving was falling behind schedule and he was asked how the work was going to get back on schedule and complete before the coming flood. His response was to construct the A-frame gantry on rails to suspend the eight pile-driving hammers. It was approved and the work completed on time. He remembered the great difficulty in driving the piling through the indurated clay. "Nobody dreamed it would be so tough!" And finally the bottom line, "Yes," he said, "the job was profitable."

Mr. Telford was not over-awed by the large turbines and generators or the huge Third Power House building. His company had been involved at the 43-acre Michoud, Louisiana plant that built the main stage of the Saturn rocket booster that carried men to the moon, so I have read. But that great, white sheet of water over the dam spillway did impress him—as it does all.

Timber bulkheads in place over two slots (blocks 36 and 38) forces river through blocks 32 and 34. Concrete placing continued on other blocks delayed for diversion of the 1937 flood. Pumping stations for job water supply now located on face of dam blocks 39 and 40. October 29, 1937.

Spillway bucket being readied for flooding. Trestle legs removed from face of dam and recesses filled with concrete. November 23, 1937.

Area within downstream cofferdam now flooded and the cofferdam is being removed down to original riverbed. River now diverted through only two diversion slots. Dust in right foreground from west mix during transfer of cement to mix plant. November 26, 1937.

River diverted through blocks 32 and 34. Timber bulkhead gates over slots in blocks 36 and 38 protected the placing of concrete as those blocks were raised to the next level for the succeeding diversion. Large steam-powered crane at left center used to service the work here after the downstream trestle was removed. November 24, 1937.

The final stages of the diversion and care of the river by MWAK consisted of closing off the slots in the dam through which the river flowed during the low period in the fall and early winter of 1937-1938. Stoplogs were dropped into slots in the blocks adjacent to stop the flow before large, temporary gates were towed to the openings at the downstream and the upstream ends of the slots permitting dewatering. Finally the slots were filled with concrete as the contract work was brought to a finish on January 10, 1938. Thereafter the river flowed over constructed portions of the dam.

Diversion and care of the river during the CBI contract was accomplished using eight large, steel gates and nine roller gate tracks. The latter were lowered into place resting on the upstream face of the adjoining blocks and the low block through which the water was flowing. The gate was then lowered to permit construction. In this manner, the river was shifted forth and back. When sufficient height of the dam was obtained and the reservoir raised from the obstruction, all of the water was carried through the outlet tubes unless inflow exceeded outlet capacity.

There was a lot of horsepower racing along in that river and the chief steward, who was in charge of the gate setting, and his river jockies (the foreman and his crew) down on the floating rigs handling the gates into and out of position had some of the most dangerous and important work on the entire undertaking. And they did it with only minor mishaps. I regret that I did not know them.

With the bulkhead gates now over the slot in block 34 the river level above the dam rose and started flowing over the remaining low blocks all across the spillway section of the dam. Crane at tower base removing cofferdam cribs in blocks 39 and 40. December 21, 1937.

CBI removed the trestle (MWAK) at elevation 1024 and engaged Bethlehem Steel Co. to build another at elevation 1180 to serve the remainder of the dam construction. It was similar in design to the former. The 1938 flood passed over the partially completed dam, as shown. July 9, 1938.

CBI selected steel gate designs for its means of closing the slots through which the river flowed. Such gates shown here in place in blocks 34, 36, and 38. Closure gates were no longer required over the downstream ends of the slots which had been constructed above the tailwater level. October 13, 1938.

Steam-powered A-frame hoist mounted on barge placing closure gate over slot in dam. February 13, 1939.

Work in the spillway section of the dam continued through the winter. In order for the concrete to harden more rapidly, the forms and the exposed face below certain blocks next to be used for diversion of the rising waters were enclosed and the enclosure heated. This precaution reduced the possibility of damage from the flowing water. Note the covers over portions of central section of spillway. January 23, 1939.

Within a month, (see photo dated January 23, 1939 this page), water is flowing over the just-completed concrete. The rising water now diverted over remaining low blocks. February 22, 1939.

With the increased stream flow additional channels were required for the passage of the river without flooding the works in adjacent blocks. April 23, 1939.

As the flood peak approached in 1939 the river was carried over 15 diversion slots. Concrete work then was concentrated on the abutment sections and on five of the spillway blocks containing the 1036 elevation outlet gates. June 1, 1939.

Work continued adjacent the rushing streams. The outlets were lined with steel curved to deflect the discharge downward into the spillway bucket. July 7, 1939.

When the flood receded and the outlets at elevation 934 were operable, the river was diverted through fewer channels. August 22, 1939.

By late September 1939 the entire flow of the river passed through the lower outlets while work continued across the entire expanse of the dam. September 25, 1939.

When the lake behind the dam rose above the middle set of outlets they too were brought into use. Construction of the pumping plant dam at the far right also was kept ahead of the rising lake level. April 21, 1940.

The increased river flow in the spring caused the water in the reservoir to rise—thus increasing the discharge through the outlets. April 21, 1940.

By early June the lake had risen sufficiently to overtop all of the slots planned for use in 1940 diversion. By now the left powerhouse structure had been roofed and the window openings were closed. June 8, 1940.

By early August 1940 the concrete in all of the blocks of the dam had been placed to levels above the 1180 trestle. At front center note the six concrete deflectors that shielded the trestle legs adjacent the three lowest diversion slots for the passage of the 1940 flood waters. Reinforcing steel rising to the skyline is for the drum gate chambers and anchorages. August 7, 1940.

After the 1940 flood season all of the blocks in the spillway section were constructed to the drumgate foundations and the river was diverted through the outlet works. The 1941 flood was passed through the outlets. At the far end of the dam the pumping plant wingdam, and trash rack structures for the 12 pump inlets nearing completion, in the 18 main unit power penstock trash rack structures the steel trashracks (dark lines) are being installed for the west power plant. At mid-section the trash racks for the outlet gates were in place and the guides being installed. The spillway bridge piers—between the drum gates—nearing completion. The two dark areas at right (top of dam) are twist slots. March 3, 1941.

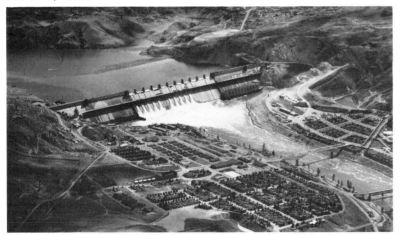

By mid-year, 1941, the drum gates were in their chambers with the crest at elevation 1260 ready for the flood—but the 40 outlets (upper and middle tiers) held the flood (the lowest of record) so no water flowed over the gates that season. June 15, 1941.

RESERVOIR ACQUISITION AND CLEARING

The reservoir behind Grand Coulee Dam extends 151 miles upstream to the Canadian border at the maximum water level of 1290 feet above sea level. In order to control and protect the reservoir and the public, decision was made to acquire all of the lands adjacent the Columbia River and its tributaries up to eleveation 1310, which approximates the top of the dam. Surveys were made of the lands and improvements and relocations of the roads and railroads included in the "taking line" began. The property owners of the communities of Kettle Falls, Keller, Daisy, Fruitland, Inchelium, Miles, Hunters, Lincoln, Cedonia, Gifford, Rice, Marcus, Evans and Boyd were compensated for their properties and they then relocated to higher ground while the small settlements of Plumb, and others were not re-established. Graves were moved and identified—some with markers identifying the remains as "unknown." The land was acquired in fee title and the reservoir was cleared of all improvements, trees and brush. Bridges remained until relocations permitted removal, and some of those steel structures were loaded onto large barges with the rising reservoir and removed to the dam site for salvage at the beginning of World War II. Clearing operations were performed with W.P.A. personnel operating from fixed and floating camps along the river and reservoir. The road and highway and railroad relocations were constructed by contractors. Steel bridges were constructed across the reservoir at Kettle Falls and over the Kettle River arm for the railroad and for the highway (U.S. Route #395) over the reservoir at Kettle Falls and over the Spokane River arm near Miles (S.R. #22). Crossing of the reservoir by motor powered ferry was provided by the state (S.R. #4) at the mouth of the San Poil River (Keller Ferry), also the private ferry at Gifford-Inchelium continued.

The lands on the right bank of the river lay in the Colville Indian Reservation and some of that in the right bank of the Spokane River lay in the Spokane Indian Reservation. The Secretary of the Interior was the trustee for those tribal lands and acquisition en blanc, was obtained for desired flowage rights over those lands. Private holdings were secured in fee title.

Survey group at Kettle Falls. Leo Ladenpara, Gene Rinehart, Larry Willouby, Dale Johnson, Stan Bishop, J.E. "Ernie" Hill (party chief), Boots Harness, Oliver Watson, Fred Saalbach (party chief), Dick Greene, Floyd Lyons, Jimmie Smith _____, Chuck Wells, Tommie Grier, Trig Hansen, Art Swanson, Althe J. Thomas (locating engineer). Photo courtesy of Tommie Grier. August 18, 1937.

Phil Nalder (seated right), and his Right-of-Way crew. R.G. Vernon, Don Williams, Gordon Whittaker, Jim Guistino, Deur Johnson, George W. Howe, Tommie Mutch, Thoralf Torkelson seated at left. October 28, 1937.

FILLING AND OPERATING THE RESERVOIR

The contractor's construction program and his means for care of the river resulted in the gradual raising of the level of the reservoir over a period of years beginning in 1937 and finishing in 1942. With the saturation of the river and the reservoir banks sliding and subsidence of the land occurred in many places and slides are still occurring in 1986. The top of the slope of the moving ground moved beyond the acquired land boundary and many additional surveys and acquisitions followed. The large slides in the reservoir banks caused huge waves to travel along the water surface and run up on the shorelands, sometimes at considerable distance from the source.

The Hawk Creek slide was one vivid in my memory! My son, then age 13, and I planned a boating venture on Lake Roosevelt as a weekend outing. The boat was small and the motor was very small. I had mentioned to the project geologist "Brownie" Walcott, that I planned the trip up the lake that weekend and asked if there were any good places to camp, handy to the lake and what features might be of particular importance. He told me that there was a nice camp spot on the shore at Hawk Creek, and while there, the slide nearby, which had

occurred two or three days previously, might be of interest. He had not yet gone to see it. It was a beautiful day and the water calm—hardly a ripple—and we didn't find a suitable spot to stop for the night as dusk closed in upon us, so we pushed on, hoping to camp at Hawk Creek. When we arrived, it was a black night with no moon and with only a meager light against the steep banks, it was even darker. There was much turbidity to the water there, and we found no trace of the creek flowing into the reservoir—nor could we find any campsite. It being nearly 11 p.m., we decided to just sleep in the boat, tied up to a snag protruding from the water and covered up for the night.

But our sleep was fitful and interrupted by the splashing of the "large carp feeding in the water." As daylight approached and the skyline appeared steeply overhead, I realized that we were in a very dangerous location. What we had thought were carp playing in the water was the splashing of masses of earth falling into the water as the face of the sliding mass continued into the reservoir. With great haste, we departed as fast as the means at hand would permit! That slide, earlier in the week, had suddenly dropped several million cubic yards of earth about 300 feet into the reservoir at the right bank of the Hawk Creek mouth. The slide moved directly toward the left bank of the creek

The author, L.V. Downs, president of the "Paul Bunyan Club" at the time, and Mr. F.A. Banks at the launching of the good ship *Paul Bunyan*. It was the work boat on the reservoir clearing activity. January 4, 1939.

Paul Bunyan—the work boat on Reservoir. 1939.

mouth with such momentum that the ensuing wave ran up the bank, breaking off 10- to 12-inch diameter pine trees 60 feet above the reservoir level. As the wave continued down the left bank of the reservoir saw logs were beached about 15 feet above the highwater mark about 2 miles downstream. In the construction and in the operation of the reservoir, it was most fortunate that it was in a very sparsely-settled area and that water-based recreation had not yet become a very affordable popular pursuit. I am not aware of any loss of life on the shore lands or surface of the reservoir because of the unstable banks, but we did receive calls from Roosevelt Lake Log Owners Association and other upriver interests, concerned about the short-term fluctuations of the reservoir.

After the contractor's work was finished and operation of the reservoir became a reality, boating became popular. To prevent boats from endangerment, a log boom was constructed across the reservoir about ½ mile above the spillway. It was a public safety measure, as Mr. Banks, the project construction engineer described it, ''to prevent some unlucky boater from enjoying the last thrill of his lifetime as he sailed over the spillway and into the bucket at Grand Coulee Dam!''

The Columbia River had always been an artery for movement of floating debris of all kinds down to the sea and that continued for awhile after the dam was built. The protective boom at the dam was not sufficient to hold the large masses of driftage at flood time, so the boom was opened, passing the material over the spillway. But it did not go on immediately down the river—eddy currents carried much of it back into the tailraces against the powerhouse walls and adjacent the dam in the spillway. The abrasion of the concrete from such material could not be permitted to continue and efforts to route or tow the floating heavy logs, trees, etc., into the stream channel were not successful. Such efforts were expensive and merely passed the problem on to each dam in succession down the river. Men who had worked on the river and were familiar with its quirks, discouraged us from attempting to catch the floating material at sites upstream from the dam where it could be beached, salvaged or burned. But we experimented with small test booms in the west canal of the Columbia Basin Project, near Ephrata, satisfying the idea and hopes that booms could be kept intact and effective in fast water, if the shear angle was sufficient.

An agreement for participation was negotiated with all power dam operators on the Columbia River and a boom facility was constructed at China Bar, near the head of the reservoir. There, the annual catch was beached and burned, keeping the entire river and its many reservoirs relatively free from floating hazards to the dams and appurtenances and to the public. Now, since 1975, barge-mounted burners are used at China Bar and in the Kettle River to dispose of the unwanted flotsam.

TAILRACES AND RIVERBANKS

The tailraces for the power plants are the excavated channels extending from the far ends of the plants out to the river channel. The channels carry the water coming from the turbines into the river, so are constructed to provide adequate capacity without causing the water to "pile up" causing a reduction in the net energy that can be produced. Also, the velocity of the water from the turbines coupled with the turbulence, wave action and surging caused by the spillway discharge required extensive protection of the earth slopes after excavation had been completed. Such protection also was provided along the riverbanks beyond the tailraces. Thick blankets of rock (called rip-rap) from the excavations were placed to cover the slopes of the channels. During the period of construction, numerous slides occurred and the excavated slopes had to be flattened repetitively. To assist in stabilizing these slopes, shafts were excavated, extending to bedrock and pumping of water, seeping through the clay overburden, was commenced in a continuing effort to avoid further slides and instability of the ground on which the government camp and access roads are built. Then came the great flood of 1948, when a flow of 637,800 cubic feet of water per second occurred at the peak of the flood crest. Extensive displacement of the rock rip-rap, along the riverbank occurred for about a mile downstream. Where all of the protective cover was dislodged, temporary protection was provided by dumping rock from spoil banks to restore those slopes. After the flood, J.A. Terteling Co. was engaged to repair the damage and to quarry very large rocks from the granite outcrop just north of the town and place the material as a protective cover, strengthening the resistance against future floods. A correlation was observed between the rapid drawdown of the river and the timing and severity of ensuing slides such that limits were placed on the rate of change in river levels in order to forestall such ground instability. When pertinent data was being prepared for the design of the Third Power Plant in the mid-1960s, recommendations included removal of the row of residences along the river on the west bank with substantial slope unloading. But that option was passed up, favoring stabilization by less drastic attempts to cure the problem. Thereafter, continuing efforts to combat the erosion and subsidence of the riverbanks not only in the near vicinity of the dam but for several miles beyond, were begun. The problem was worsened by the fluctuations inherent in the greatly increased hydraulic discharge from the additional power plant. The riverbank stability is an ongoing problem even now in 1986, as the river continues its work of taking the land down to sea level.

MIGRATORY FISH

The Columbia River was a principal route for spawning salmon and steelhead trout migrating to headwater streams. The preservation of this resource was certainly an objective in the planning and design of the Grand Coulee Dam project. Because of the great height of the dam, it was considered infeasible to construct fish ladders over the dam. Instead, large fish rearing facilities were constructed on the Icicle River near Leavenworth, with smaller facilities on the Methow River

Fish ladders were provided and used during the 1937-38 fish migrations. Shown here is ladder in block 62 (the left slot in photo). January 27, 1938.

near Winthrop, and near Ford on a branch of the Spokane River. These facilities were then used for salmon and trout production. Adult salmon were trapped at Rock Island Dam then transported to the Leavenworth hatchery, held in ponds until ripe before being artificially spawned and the eggs hatched. The fry were then fed in the waters of the facilities and the salmon and steelhead fry were released into streams entering below Grand Coulee Dam. Since the rearing facilities were not in readiness by the summer of 1938, it was necessary to provide temporary fish ladders in two of the blocks (numbers 38 and 62) of the dam to permit the migrating fish means for reaching their traditional spawning areas. This facility was used in the 1937-1938 runs only. There I saw migrating salmon and steelhead jump the ladders, but the lamphreys climbed the vertical walls of the slots. With the salmon run over for the season on October 31, 1938, the contractor was free to proceed with construction.

The facilities provided did protect and perpetuate the runs as indicated by the number of salmon counted at Rock Island Dam in the years before and after the construction of Grand Coulee Dam. In view of the controversy that has occurred, associated with the care of the fish runs because of the construction of additional dams on the Columbia the following may indicate that Grand Coulee Dam with its propagation facilities and restocking lower tributaries did not kill off the salmon resource in the river, but apparently enhanced it. Steelhead trout did not fare as well.

FISH RUNS COUNTED AT ROCK ISLAND DAM

Species	1933-1946 Aver.	1947	%
Chinook	7,830	11,727	150
Silver	66	229	347
Blueback	17,357	79,498	458
Steelhead	3,058	1,963	64
TOTAL	28,311	93,417	330

FLOOD CONTROL

Flood control on the Columbia River is a responsibility of the Corps of Engineers. After the 1948 flood the forecasting of probable flood flows received greater emphasis and the reservoirs on the river and its tributaries were drawn down each spring in anticipation of the flood

Grand Coulee Dam during the big flood. May 30, 1948.

peak flows. With the Canadian Treaty and the Columbia Storage Power Exchange agreement, many more entities became parties to the flood control activity. The Corps of Engineers is still the principal agency in the flood control operation, but flood control and power operations are coordinated. The River Forecast Center of the National Weather Service in Portland, Oregon, prepares the annual flood forecasts and monitors precipitation, temperature and runoff data and updates the forecasts periodically. The Bureau of Reclamation, Bonneville Power Administration—it also represents British Columbia—and the Coordinating Contractors—power purchasers of Canadian entitlement—all are affected by the activity and maintain an active interest in the flood control program. The Coordinating contractors include the Public Utility Districts of Grant, Chelan, Douglas, Clackamas, Cowlitz, Pend Oreille and Skykomish Counties; Seattle and Tacoma, Washington; Eugene Oregon Power and Light Co.; and the private utilities, Washington Water Power, Pacific Power and Light, Portland General Electric, Puget Sound Power and Light and Montana Power Co. The Corps determines the amount of storage to be drawn from each reservoir on the entire river system according to the storage levels and the forecast inflows and notifies the parties as to the timing and duration of the releases.

CHAPTER V
GETTING DOWN TO THE BOTTOM OF IT

EXCAVATION — COMMON

The contract required MWAK Co. to dispose of all excavated materials not required for permanent features to be disposed of at a distance not less than 3500 feet upstream from the dam, but not in the river. The contractor set his plan on the deep side channel "Rattlesnake Canyon" on the left side of the river and to haul the excess from the right bank excavation for disposal on the wide bank of the river upstream of the damsite. The major portion of the material to be excavated was on the left bank and the contractor built a great, 60-inch-wide conveyor belt (Jeffery Mfg. Co.) system, serving both sides of the river onto which excavated earth was fed through feeders or hoppers into which the material had been dumped from trucks and tractor-pulled, self-dumping wagons which, in turn, had been loaded by the shovels operating at the face of the excavation.

At the outset, beginning in December 1934, attempts were made to pass heavier boulders found in the excavation onto the belt than it was capable of handling without damage, but after grillages of suitable openings on feeders were used, the system proved to be speedy and efficient. Because of the distance (1 mile) and elevation (500 feet) through which the material had to be transported, the system required numerous flights or runs of continuous belts, each discharging onto the one farther from the source. This conveyor system was designed to transport 60,000 cubic yards of material per day and in July 1935 averaged 55,514 cubic yards. It had motor capacity of 5,000 h.p. With this conveyor system a total of 10,688,500 cubic yards of earth was transported in the 23 months it was in service—it was a "river of dirt." At point of disposal a stacker—so-called—was used to distribute the waste over a wide front as the fill enlarged. The clay materials as deposited advanced, maintained a face slope of almost 1+ to 1 for a height of about 40 feet, but the entire mass periodically slumped to a slope of about 10 to 1 and the stacker was toppled more than once.

Several kinds of tractors and hauling vehicles were used—here are Athey wagons (on tracks) and Woolridge wagons (on pheumatic tires). Equipment manufacturers cooperated in developing improved excavating equipment. I am told that Caterpillar Tractor Co.'s first diesel engine-powered tractors were used here. August 1935.

View of belt conveyor system extending from the feeder grizzly at left of crawler tractor dozer across the river and up the far canyon wall to point of disposal. Transfer stations where break in slope appears were used to limit length of belt sections and to permit distribution of loadings on motor drives. On steep grades the spacing was reduced. October 29, 1935.

Earth covering bedrock was finally removed by hand shovels before drilling of the blast holes began. Typically drill crews drilled out sufficient holes for the powder men "powder monkeys" to load and blast each shift. April 29, 1936.

At the end of a shift the area would be cleared for a safe (?) distance and the blasting then occurred. Dynamite in sticks and primers in which electric blasting caps were inserted were tamped into the drilled holes then the wires were extended to a source of electricity for the blast. August 1935.

In the narrow, deep trenches in the bedrock the drillers worked in a cloud of dust—blowing the cuttings from the holes by an air pipe forced to the bottom of the hole. Wet drilling was not practiced or required in those years. Tony Falbo and crew. March 26, 1937.

100

Manufacturers of heavy excavation equipment vied for this big job and efficiency tests were used to field-try various tractors, dozers and wagons in the contractor's search for the best. The overburden was removed in lifts so that the excavated slopes could be controlled and the material safely removed without unstabilizing the mass. Nevertheless, while excavation was underway, in the pit on the right bank slope instability developed and work had to be suspended to protect the cofferdam near the top of the sliding material. At the toe of that sliding material was a narrow, deep channel in the foundation rock which was only excavated after the unstable material had been stabilized by freezing what, in reality, was a frozen earth arch dam across the channel. Fortunately the method worked, permitting the work in the channel to proceed. The ice dam was maintained intact until that channel had been excavated and the concrete of the dam brought up above the level of the frozen structure. The power shovels for the major excavation task were new, electrically-powered crawler types of 3- and 5-yard capacity. Smaller gas-powered shovels and draglines were used in confined, steep or uneven ground. Much larger shovels and trucks and large front-end loaders were in the offing but they had not yet been manufactured.

In hindsight it appears that the contractor looked with mixed feelings at the occurrence of the slides. They did induce some added hazard to the equipment and workmen but also involved additional excavation and compensation. The contractor protested the manner of payment for the slide above the ice dam and filed a claim. In testimony under oath before the Court of Claims, one of the witnesses stated that when Mr. Guy F. Atkinson appeared on the job the morning after the big slide, "He just rubbed his hands together and said, 'Goody, goody!' "

EXCAVATION — ROCK

Because of the importance of this dam, 36-inch diameter holes were drilled into the bedrock to ascertain the undisturbed condition of the rock at depth. The holes were entered for visual inspection and the large cores were also available to view. Tests of the granite showed compressive strengths from 10,000 to 34,000 pounds per square inch with average tests 22,000. Rock excavation after removal of the overburden was carried to a depth of about 3 feet as a minimum and to

such additional depth as necessary to remove weathered or rock of doubtful character for the foundation. Jackhammers with steel shafts or rods and forged bits were used by MWAK for drilling blasting holes. Later, removable bits came into general use. Each shift, holes would be drilled, the holes loaded with dynamite and blasted as the shift ended. Excavation followed promptly. Final preparation of the rock involved the removal of loose and unsuitable rock by barring and wedging before sandblasting and washing preparatory to placement of concrete.

In two channels in the rock, the excavation was carried to some additional depth without getting into suitable rock whereupon in order that the work could proceed, shafts were excavated and extended further into the formation and then filled with concrete. The shaft in Block 40 was formed in the concrete, extending into the lower gallery of the dam, permitting removal of the weaker material and replacement with concrete as the larger scale concreting continued.

TREATMENT OF FOUNDATION ROCK

When dams are built, the impounded water under increased pressure at the base of the dam tends to force its way into the naturally occurring cracks and seams in foundation rock under the dam and to migrate downstream where the pressure is lessened. To reduce this percolating water migration in the foundation rock under the upstream portion of the dam foundation, the rock is consolidated and cracks and open seams are filled in a process called grouting. A mixture of cement and water is forced under pressure into the rock through holes drilled for that purpose. On this dam, the grouting was done in three stages. The first was through holes drilled 30 feet into the rock along five lines at 20-foot spacing and with the holes at 20-foot spacing. This "blanket grouting" was located under the upstream portion of the structure and the grouting was accomplished at a maximum pressure of 200 pounds per square inch. After the concrete had been placed and hardened, but before the area was flooded, additional holes were drilled on an incline sloping under the dam through pipes embedded in the concrete and exposed in the face of the upstream fillet of the dam near the contact with the unexcavated rock. Those holes were deeper—75 feet—and were grouted typically with thinner mixtures of grout and under higher pressure, 300 pounds per square inch. A third grouting sequence was then used when construction of the dam had advanced appreciably through holes drilled through pipes which had been placed into the rock and extended into the lowest gallery of the dam at the time of construction. These holes spaced at 10 feet alternately 150 and 200 feet in depth, sloped slightly upstream and were grouted from the gallery access using maximum pressures of 400 pounds per square inch. Thereafter, drainage holes were drilled at 10-foot spacing, also from the lower gallery after all the grouting was finished. These holes of 3-inch diameter and 50-feet deep drilled through downstream dipping pipes embedded in the concrete were connected into the drainage gutter—their purpose being to permit the escape of any water percolating through under the foundation of the dam.

At the time Grand Coulee Dam was constructed, pressure grouting techniques were still somewhat in the development stage though extensive grouting had been performed at Boulder/Hoover Dam. Grout under high pressure is very abrasive to materials it contacts and it tends to deposit on the surface it contacts. The contractor made numerous improvements in the grouting process, principally in high pressure pumps where replaceable hard surfaced liners and piston rods and rubber-jacketed pistons were used to maintain the serviceability of the equipment. Grout admixtures and chemical grouting and other improvements were still ''Not yet.''

At top of photo, gravel from the pit to the raw storage pile moves through the primary crusher, a secondary crusher and into the gravel screening plant for wet screening and separation into desired sizes. May 6, 1936.

Gravel plant showing conveyors delivering processed and sized sand and coarse aggregate to the live storage piles from whence delivery was made to the mixing plants on order from that point by remote control. To avoid breakage and seggregation, the processed materials were passed down ''rock ladders.'' September 8, 1937.

CHAPTER VI
CONCRETE FOR THE STRUCTURE

THE AGGREGATE

Aggregate for use in the construction of the dam was a major item in consideration of its construction. Sizeable deposits of good quality aggregate were found to exist at a number of locations along the river so selection of the overall most suitable source or sources became the focus. A good prospect for aggregate was first investigated in 1933 at "the Plumb pit" located some miles up the river above the dam, then in 1934 deep test pits were dug by hand testpitting methods in the "Brett" pit, a deep, large remnant glacial deposit high on the right bank of the river canyon. It proved to be adequate as to quality and quantity but included a great excess of sand which would require selective use and wasting large amounts of the fine materials. The latter resulted in there now being a 12,000,000 cubic yard sandpile at the town's back door. As much sand was included in the mix as was prudent keeping in mind that the more sand being used over the optimum the more cement was needed to obtain the optimum strength. Within the deposit was also very fine rock flour from the glaciation. It proved to be very beneficial and seemed to serve the same or similar beneficial function as so called "pozzolons." In fact, pits having pozzolitic materials were investigated in the project area but not used except for a number of test batches.

CEMENT

Cement for the structure got early attention also. The laboratory of the chief engineer of the U.S.B.R. in Denver had, in cooperation with the various cement manufacturers of the west, been striving for a better product seeking to, among other characteristics, reduce the generation of heat in the setting of the cement in the large structures as well as to increase the strength of the concrete. Also, attention was directed to the great quantities of cement that would be needed should Grand Coulee Dam become a reality. Of course, the low ebb of the economy made increasing cement or other manufacturing capacity optimistic financial ventures. One of the improvements made was the use of so-called "low heat" cement, which reduced the temperature rise in the concrete—one of the trade-offs being slower setting and the need to leave forms in place longer before removal. Transport of the cement was in bulk in the then-standard wooden boxcars. Bulkheads were placed across the cars before loading so the doors could be opened without loss of contents. Unloading at the bulk storage plant along the railroad at the approach to the damsite was by large, electrically-powered "vacuum cleaners." These machines were controlled in movement and operation using hand-held "wands" wherein mercury switches responded to the mere tilting of the wand moving the machines forward, back, right or left. At the steel-tanked bulk storage plant the cement from the several mills was blended for uniformity and aeriated and transported to the mixing plants through 11-inch diameter steel pipelines using compressed air. When aeriated, dry cement flows almost like water. At the peak of construction activity, 90 cars of cement were used in a day and on one day 125 cars were unloaded. Industry began to focus on more efficient transport and the use of hopper-bottomed tank cars for not only cement but many other bulk materials became standard practice some years after the dam was built. Cement for the grouting needs was delivered in cloth bags closed by a wire tie at the open end. A sack of cement, one cubic foot in volume, weighed 94 pounds, and was handled, opened and shaken by hand. The empty sacks were returned to the manufacturer for reuse. The use of paper or plastic for cement sacks was yet to come. Over 11,600,000 barrels of cement were used in constructing Grand Coulee Dam and related features by the close of 1943.

Because of CBI's planned rate of concrete production, the output from four additional mills (total 9), was required; three in California and one in Montana. Both modified cement for powerhouse and other structures and low heat and modified cement for mass concrete were required.

BATCHING THE CONCRETE

The concrete for the dam was specified to be the best for the money at the time. Good, uniform concrete was desired with adequate strength, durability and economy. The concrete would be made of aggregate from nearby sources, cement and water. No admixtures for inducing special properties other than calcium chloride for accelerated set, had yet become acceptable. Economic considerations focused on a good grading, large-sized aggregate, minimum cement, effective operations and ease of handling. Where exposed to freezing and thawing and abrasion of running water and other abrasive forces, wear-resistant aggregate was needed and denseness and homogeneity of the concrete essential. For these needs, the Brett pit materials were excellent. After mixing, the concrete had to be placed with care to preserve its quality, to avoid segregation, to consolidate it, to avoid separation or discontinuity between successive layers, to prevent impairment after placing and to insure its adequate curing to prevent water loss until it was cured. These considerations required a low water/cement ratio and a well-graded sand without an excess of fines; the separation of the coarse aggregates into several sizes and to prevent damage or contamination prior to mixing; dependable equipment; effective methods; automatic control of operations; effective organization; and quality control of the product. The actual proportions of the materials in the mix were varied somewhat to meet the output from the gravel pit because the several sizes needed for an optimum grading was seldom found day to day in the deposit. However, the product was maintained at high quailty. But the excess sand could not be brought into the mix. Maximum size of aggregate used would pass through a screen with 6-inch openings. Larger material was crushed and used. MWAK and CBI met these requirements with a vigorous approach and planned, designed and constructed, operated and maintained the needed works, employing competent personnel.

Raw material from the pit excavation was routed through a heavy grid with 16-inch openings passing material going through crushers to the raw storage pile.

The large rocks not passing the grid or grillage of railroad iron was hauled to the edge of the pit and passed through a very large, 36-inch as I recall, jaw crusher then back into the raw storage, or surge pile. Next the material was wet screened and sorted, excess sand removed and transported by belt conveyor to the waste pile and four sizes of coarse aggregate plus sand stored in separate adjacent piles to provide reserves for periods of interruption and variations in pit yield prior to use.

The remaining sand was separated into three sizes by a classifier and deposited into three separate piles. The inspector periodically took samples of the sand from each pile and from tests thereon determined the proper proportions of the three sizes desired. He then set the controls on the variable speed feed belts drawing sand from each pile. The recombined sand was then run through a "beater" which mixed it.

The aggregates so processed were not permitted to be dropped from excessive heights to avoid shattering. As required for concrete, the sand and separate sizes of other aggregate were drawn from the piles through gates, dropping it onto a conveyor belt on which it traveled to the appropriate bin in one of the concrete mixing plants. Conveyor belts were used extensively in the handling and processing of the aggregates. It was an ingenious plant of large scale, and its output had to be kept in close harmony with the demands of the concrete mixing and placing since both the raw storage piles and the piles of processed aggregate were limited in size and all had to be available of the proper sizes and proportions for the mixing needs. The operation of conveyors and gates were controlled with electric controls by a bin operator. There, buttons were pushed to keep the separate bins supplied, routing each size into its bin before the next size was called up. Cement bins were kept full by telephone order to cement pump operators at the silos.

The contractor's two mixing plants then often referred to as "the house of magic," were for their times, just such. Those plants were actually large, multi-storied manufacturing plants set up on the canyon walls adjacent to the dam, wherein the two grades of cement, sand and four sizes of coarse aggregate and the water were each received, stored and replenished to permit continuous operation. From the bins and tanks, the materials for each batch were drawn into separate scale bins which were automatically controlled to shut off when the exact bin weight was reached. As soon as all materials were weighed and the discharge chute directed into the proper concrete mixer, the batch plant operator tripped the batch in the desired sequence needed for most efficient mixing. The scales for each of the ingredients in the concrete had multiple beams which were pre-set to fit the proportions of several different mixes that were used regularly so that the operator could merely select the appropriate mix (though over 50 separate mixes had been formulated by "Concrete Control" to optimise the concrete with the yield of the aggregate pit) called for by use of a 5-point selector switch. The weights of the individual ingredients for each mix was recorded on charts in a panel at the batch plant operator's station.

A "dispatcher" received the orders for concrete from the foremen of the concrete placing crews by telephone and relayed the orders to the batcher operator by signal systems. He also observed the train movements and identified the special mixes to the hook tenders by tags as those mixes were dumped into the buckets. He communicated with the mixing plant operator by speaking tube. At the peak of production, with 20,684 cubic yards placed in 24 hours of May 29, 1939, each plant was loading onto the trains 108 batches per hour or one every 1.8 minutes. A remarkable achievement of efficient plant, organization, supervision and skilled operators! The noise level in the plants was very high and the constant demands for alertness necessitated short, frequent rest "from the grind." Sufficient operating personnel in the plant permitted rotation of duty so that two could be at the rest station concurrently. The importance of these plants to the contractor's success cannot be exaggerated—when the plants were idle for any reason, no income from "concrete in the dam" was accruing and there would be a lot of equipment idle—and idle manpower also.

MIXING THE CONCRETE

The USBR and contractors' personnel—also the equipment manufacturers—had learned much about the mixing of concrete on large projects by the time Grand Coulee Dam got onto the drawing boards. By then, Owyhee Dam was complete and Boulder/Hoover was abuilding. Also the Bureau's chief engineer had established an engineering laboratory in Denver for research on cement and the mixing and testing of concrete. But testing the strength of mass concrete with large aggregate therein, as being used in dams, had not been done. To permit such testing, the Bureau installed a 5,000,000-pound capacity testing machine in its Denver laboratory at the U.S. Custom House. There mixes using aggregate from the Grand Coulee Dam sources as well as from other dams under contemplation or construction were tested for strength using cylinders up to 36 inches in diameter and 6 feet in height. When the first of these large cylinders was tested to failure, it shook the building—we thought it was an earthquake. While determining mix efficiencies as to proportions, grading, water to cement ratios and slumps continued in the laboratories, the actual efficiency of the mixers out on the job seemed to warrant further attention. The 2½ minutes specified as the

minimum time for mixing any batch of 4 cubic yards of concrete at Grand Coulee Dam did not guarantee that with the mixers procured by the contractor, the concrete would be homogenious or of optimum quality, nor was it! Manufacturers of such large equipment also did not have all of the answers in the search for greatest efficiency and economy of the mixing task. Speed of rotation, size and number of baffles or blades, their position, alignment and clearance from the drum wall all may bear on the time required to secure thorough mixing even with most effective charging sequence. As the work on Boulder/Hoover Dam neared the end, Mr. Orin G. Patch was transferred to the Grand Coulee Dam project on July 27, 1934 and was assigned the responsibility for continuing investigations and selection of the concrete materials, for the field laboratory and for the inspection of the materials and mixing of the concrete. Mr. Patch was well suited by experience, ability and temperament for the job. In my judgment, he and the engineers and technicians working under his supervision are most deserving of credits which are due for the outstanding results obtained in the quality control and uniformly high consistency of the concrete as manufactured. In so saying, I do not intend to negate the role of the chief engineer's laboratory, of the manufacturers and the contractor's engineers, technicians and supervisors—the contractor was very cooperative in seeking improved efficiency. But those had a profit motive! Mr. Patch was primarily concerned with quality and efficiency. He had patience, perseverance, a scientific mind and he sought solutions and answers or explanations* and he had good help.

When MWAK mixing plants were operable a series of tests designed and initiated by Mr. Patch and his staff were conducted to

---

* A case in point that I recall—a new, small mixer was received and put to use mixing concrete of various mixes for test purposes. The concrete was cast into cylinders for use in various investigations. It was noted that the test cylinders cast from the first mixes from the new mixer did not shrink as expected but expanded in height after casting. Later, mixes from that equipment expanded less and finally produced concrete having usual shrinkage characteristics. The mixer interior had been painted with aluminum paint and as it wore off during the mixing, added its effect to the product. Mr. Patch recognized the value of this powdered aluminum as an additive for some purposes where non-shrink concrete was desired. And, he determined the very small proportion to be added to the cement to get the degree of shrinkage or expansion desired.

increase the efficiency of the mixers because "thorough mixing" required up to 4 minutes or more. Better consistency and more uniformity and reduction of mixing time all were sought.

Soon after concrete placing resumed in 1936, the results of mixer efficiency tests conducted in the local concrete control laboratory and on the job permitted the reduction of the mixing time per batch from 3½ to 2 minutes when the operation was running smoothly and the consistency of the concrete in the mixer was uniform. The latter was verified by taking three samples from the batch, catching the samples directly as it was being discharged. The consistency of each batch was recorded graphically in the batcher room. The mixer efficiency tests were solicited by the contractor, MWAK, and it furnished the Bureau laboratory at the dam a model mixer of ⅟₃₀ capacity of the Koehring 4 cubic yard mixers in the plant. Blades or baffles in the mixers were re-sized, repositioned, reshaped and realigned and mixes were tested with varying combinations of the loading process. Samples of the mixed concrete were intercepted as it was dumped to determine the uniformity of the samples as respresentative of the total. With these efforts, mixer design was improved and better concrete was manufactured more economically and with reduced mixing time.

Three long blades, parallel to the axis proved to be significantly more efficient than the 6- or 9-blade design then in use. As the result of these tests by Bureau forces, the contractor replaced the blades in all eight mixers on the job by June 1, 1937.

The consistency meter gave an indication of the water in the mix and was based on the premise that the center of gravity of the mixer and contents would be centered with the mixer idle and the contents equally distributed but would move off center forward toward the mouth of the mixer in the direction of rotation. I understand that Mr. W.H. Clagett, the inventor of this device received a patent on it while employed under Mr. Patch's supervision. Too, I am informed that consistency meters were no longer available on mixers from the Koehring Co. when the Third Powerplant was built.

With the reduced mixing time required to obtain desired results, the contractor's rate of progress was materially increased with the result that at the peak of production MWAK manufactured and placed 15,844 cubic yards of concrete from the eight mixers (four in each plant) in a 24-hour period, and in August 1937 placed 377,133 cubic yards. On the succeeding contract for completion of the dam, the contractor CBI produced and placed 20,684 cubic yards in a 24-hour period and in October 1939, placed 536,264 cubic yards.

After the materials were received into the rotating concrete mixer, the mixing continued for a time, the minimum of which was specified. The time of mixing was pre-set on a delay switch so that the mixer could not be discharged prematurely. But the operator then controlled the actual discharge, assuring himself that a "bucket" was under the spout to receive it. The mixers were each of four cubic yards capacity and the "buckets" were the same. The mixers, four in number in each plant, were equally spaced about the center line of the plant, the charging chute from the batching bins rotated to each mixer in succession or separately at the control of the operator. The mixers discharged into a single open-bottomed, ungated hopper, the bottom of which was set above the trestle deck, sufficiently for deposit of the concrete into the buckets, four of which were carried on rail cars moved by 10-ton diesel locomotives.

CBI rebuilt the mixing plants into this twin structure with eight mixers of four-cubic-yard-capacity each. Actually, the housing merely enclosed the two independent mixing plants relocated to serve the job via the new high trestle (rail height 1180). May 24, 1939.

Typically both contractors constructed wooden, prefabricated standard forms suitable for reuse wherever practicable. Forms for the interior contraction joints of the dam were faced with shiplap (later lined with light-gauge sheet metal) with the keys attached. Such forms were made in well-braced panels or sections with lengths varying as to location of use. The forms accommodated the full height of a "pour," which was set at five feet in height as the standard set by the cooling system embedments as well as the interval between the horizontal keys in the longitudinal contraction joints. The base of the forms was held in place at the top of the preceding layer or lift by anchoring to threaded rods which had been embedded therein. The tops of the forms were held by anchor rods slanted down to and attached to hairpin-shaped rods embedded at the finish of the previous pour. After the concrete had hardened sufficiently, the forms were removed and raised, set in place and reattached for the next lift or pour. Forms for confining the concrete in the first pour on any part of the bedrock were built in place and anchored using ties and braces attached to the rock. Forms for surfaces exposed to water or to view were made of vertical grained tongue and grooved fir lumber—smooth, beautiful formwork! Such lumber then was readily available. The forms were constructed in the "panelyard" of the carpenter shops and moved to the point of use by truck. As an experiment only, celotex was used on a few downstream face forms to draw excess water from the concrete.

Before the dam reached full height much discussion was had about the relative economy and desirability of having concrete surfaces formed to a variety of standards of smoothness and texture. Later the chief engineer established such standards for forms and finishes for future construction.

PREPARATION FOR PLACING THE CONCRETE (CLEANUP)

Before concrete could be placed in any location the forms were placed to proper alignment and grade. The rock and/or concrete surface needed to be cleaned, all metal items and systems installed and cleaned of objectionable coatings. Piping systems to be embedded had to be inspected or tested to insure proper functioning, grout and waterstops needed to be welded to provide continuous elements to later confine the grout into proper locations during the later contraction joint grouting. The forms were oiled for ease of stripping (removal). All of such work, when finished by the contractor, was then checked by the government engineers and inspectors—forms as to alignment and grade, embedded features as to quantity and rigidity, piping unobstructed, grout and water cooling systems tight, grout stops air tight and the foundation of rock or concrete washed and adequately cleaned with all objectionable deposits and coatings removed. Approval for placement of concrete was then authorized by the inspector.

In the preparation of rock and concrete surfaces to receive concrete the contractors on the dam developed devices and methods that were very significant economic improvements. At the outset rock surfaces

Forms were removed after the concrete had hardened sufficiently to avoid damage. The forms were raised and re-set for the next pour using the A-frame and "coffin hoists" shown. Forms were cleaned of any encrustations and lightly oiled before the concrete was placed. May 29, 1939.

were carefully picked and hand-brushed using steel brushes to remove adhering deposits on unsuitable rock prior to washing the surface to clean it of accumulations.

Concrete surfaces were carefully brushed with heavy wire-brush brooms to remove the upper ¼-inch or so of the fine material and laitance that raised to the surface as a product of chemistry of the material which, if not removed, would prevent good bond to the overplaced concrete. The brushing required much hand labor and was time consuming and expensive. Then sandblasting was introduced as a means for cleaning the rock and concrete surfaces—but the operation was primitive and it too was time consuming and expensive.

The contractor MWAK delivered skiploads of sand onto the surface of a nearby block from which workmen carried it in 14-quart pails to the point of use. A hand-held device was used, consisting of a short length of pipe (perhaps ¾-inch in diameter) to which two "Y" type fittings were installed so that the device had three inlet pipes with the "Y's" pointing toward the outlet. Valves were attached to the rear "Y" legs of the device with fittings for attachment of rubber hoses connected to service lines of compressed air and water under pressure. To the other "Y" was attached a section of rubber hose (perhaps 1-inch internal diameter) about 6 to 8 feet in length. This device was then used by two workmen, one holding it and regulating the flow and mix of air and water by regulating the valves and directing the jet onto the surface while the other held the open end of the short hose to the surface of the bucket of sand. Thereby sand was introduced into the high velocity stream of air and water, providing an abrasive jet which was able to remove the deposits from rock or concrete. This was an improvement of course, but the means for delivery and use of the sand and removing it from the worksite left much to be desired. On the later contract this arrangement for sandblasting was replaced by CBI after considerable testing and experimentation. Bins were placed for delivery of sand through the deck of the trestle from which it was drawn into a high pressure hose in compressed air to the point of use where water was added by the operator to increase the effectiveness and avoid the dust. The nozzles on these devices were subject to severe wear and high alloy steels were used to make them more serviceable. This was a significant step in more efficient construction.

Waste water, sand and fine debris from construction was disposed of by pumps and in buckets or skips until the concrete in a block was raised to permit a drain system to be installed. At that point, vertical drain pipes were embedded or formed in each block. The drains dropped into a pipe sloping downward to the downstream face of the dam discharging onto the spillway face and into the bucket or into a connecting header for disposal outside the limits of the powerhouse.

Typical preparations for concrete placement on the rock foundation of the dam. After all loose and broken rock was removed, forms were constructed in place to fit the shape of the rock surface at the contraction joint locations. A grade line was set at an even five feet in elevation (see horizontal grade strip near top of form). Anchorages and struts held forms from displacement during concrete placement. Pipes for contraction joint grouting—supply and return headers—served each joint. Grout outlets—shallow metal cups connected by ½-inch pipe from supply header—were nailed to the form. Grout stop of copper or stainless steel was attached to the form and welded to form a continuous air-tight member located about 12 inches above the rock surface, also in the longitudinal contraction joints and along the face of the dam. December 30, 1935.

Here Jack Sowle holds one of the contractor's first efforts in development of pressure-fed sandblasting nozzles. May 20, 1938.

TRANSPORTATION OF CONCRETE

At the mixing plant the concrete was discharged directly from the mixer into an open-bottomed short chute below which had been spotted a 4-cubic-yard capacity bucket for transport to the pour. The buckets as shown in photographs were closed at the bottom by a piece of rubberized belting held in place by a series of rollers arranged for ease of operation at partial or fully open positions. Special mixes were identified by the dispatcher and the "hook tender" riding the "train" then put an identifying metal tag on that bucket to insure proper delivery. Each train was a single flat car equipped with bunkers into which five buckets could be placed. The engine was a small diesel-powered locomotive on standard gauge track. The trains moved rapidly to and from the cranes serving the pour so that the concrete could be placed soon after mixing.

CONCRETE PLACING

When a block was ready to receive concrete, the concrete foreman and his crew of five or more and a signalman arrived at the block along

One of the later designs of an air/water washing nozzle. Compressed air and water under high pressure was directed onto the surface to be cleaned. April 6, 1938.

with the Bureau inspector who, with the foreman, made a quick survey of the block to see that the previously approved preparations had not been disturbed. The surface was dampened with a spray of water, and the foreman ordered the grout from the mixing plant dispatcher to be followed by the concrete as needed for the block. Grout was a mortar of sand, cement and water at a consistency suitable for brushing to a depth of about ½ inch over the surface of the block in advance of the regular concrete mix to better assure a good bond with the underlying material. The concrete delivered in buckets of four cubic yards was then placed and spread in repetitive layers, one foot in thickness until the form was filled, the top surface being sloped toward its mid-point and a sump was formed to facilitate subsequent cleaning and to avoid continuous flat surfaces through the dam. The mass concrete when delivered into the forms had a slump of two inches or less and was thoroughly vibrated to consolidate the mass and to merge the new with the grout or the underlying and adjacent concrete and to insure complete contact with the forms and any embedded features. The vibrators were electrically operated and were of sufficient size and weight to require two workmen to handle and control them. Vibrators, too, were undergoing modifications at the dam to improve their effectiveness and durability. Other, lighter vibrators operated by compressed air were also used to consolidate concrete about reinforcing steel and in confined formed areas. In the typical block of the dam, the concrete was delivered directly to the block by lowering of the bucket from the trestle where it was engaged to the crane hook and lifted off the transport car. On arrival one of the workmen stepped up on the bucket rail operating the unloading gate and distributed the concrete in the block as necessary. Then he closed the gate and dismounted as the bucket was hoisted back to the transport. The process was continued until the form was filled. In confined areas and under the trestles the concrete was delivered into the forms from the buckets by use of sectional steel pipes or through rubber hoses about 10 inches in diameter, suspended from hoppers either fixed in place for the purpose, or hanging below the bucket to which it was attached as each bucket was delivered. The deposit of concrete by dropping into the

forms was prohibited in order to avoid segregation. As soon as the concrete in a block was brought to grade, workmen began installing anchorages to provide for installations to be embedded in the succeeding concrete and for anchorage of the forms for the same.

The entire operation from supervision, work force, material supplies and equipment, processing and delivery of aggregate, form building and placing, erecting of steel and other embedments, cleanup and the placing of concrete was a well-coordinated manufacturing and construction undertaking. At the peak of the activity 15,844 cubic yards of concrete were placed in one 24-hour period by MWAK which record was exceeded by CBI when 20,684 cubic yards of concrete was placed in a like period. In total, MWAK placed 4,524,209 cubic yards of concrete and CBI 6,065,663 cubic yards, for a total of 10,589,872 cubic yards of concrete in the dam, left and right powerhouses and pumping plant by 1942.

Chuck Collins, foreman right, and crew—Floyd Rohlman, William Al Johnson, Red _____, _____, signalman, _____, _____. Placing of the concrete. The four-cubic-yard buckets of concrete were dumped onto the grouted surface and vibrated into layers about 12 inches in depth—five to each 5-foot pour. The two-man vibrators shown here were typical by 1936. The signal man standing on top of the form directs by voice the crane operator in controlling the location and height of the bucket. May 19, 1936.

After the block was okayed for the next pour by the inspector, grout—the concrete mix without the coarse gravel—was spread over the surface of the slightly dampened concrete. To assure intimate contact, the grout was broomed onto the surface as shown. The cooling system piping shown here was held in place by wire tires installed at the completion of the previous pour. At this date hard hats were not universal attire—though later to be. An inspector was present on each concrete pour. Here Leroy Shively near bucket, is in control while his supervisor Albert Moser (right) talks a problem over with the concrete walker. April 4, 1938.

Workmen walking on the top of the concrete just placed wore "snow shoes" to avoid depressions in the surface while they placed the wire ties and "hair-pin" anchors, etc. which would be used to hold forms and embedments in the next pour. August 25, 1938.

SURFACE FINISH AND REPAIR OF THE CONCRETE

Flat surfaces of the concrete in the dam galleries and concrete floors in the power and pumping plants were smoothed by hand-held wooden floats and trowels unless artistic treatment was specified. Such work was performed soon after the concrete was brought to grade and screeded off to that desired level. In the galleries of the dam the floors were broomed slightly after floating to provide a less slippery surface. The edges of the gutters and the contraction joints were shaped with edging tools to round the sharp corners. The surfaces of the concrete at the downstream and upstream faces of the dam and adjacent gallery and shaft openings were smoothed with a wood float to bring the fresh concrete normal (perpendicular) to the slope of the form. Such treatment resulted in a neat appearance on the surface and reduced the tendency toward unsightly, void areas at those construction joints. Such smoothing of the surface was also practical at the top of all blocks to facilitate forming and cleanup.

Immediately after removal of forms from exposed surfaces, an inspection was made and any defective concrete exposed there was removed and replaced using a mixture of sand and cement with just enough water added to permit it to hold in place when packed into place with pneumatic hammers. Where such areas were sizeable with openings more than six inches wide, the replacement was with concrete formed as needed. The bolts holding the forms to the threaded anchors were tapered so that upon removal they left a neat hole in the surface that also was filled with dry-pack mortar.

CURING THE CONCRETE

Curing of the concrete after placement was accomplished in several ways. The purpose, of course, was to keep the concrete from drying during the setting period. If not done, the concrete would have less strength, durability and water-tightness and surface cracking would result. To meet requirements for curing the exposed surfaces were kept continuously moist for 14 days unless covered by abutting or overlying concrete. At the outset when concrete was being placed in weather temperatures below freezing in December 1935, the concrete surfaces were flooded with water and steam lines were passed through to keep the water from freezing and warm water was sprayed onto the formed surfaces after the forms were removed but the contractor soon stopped concrete placement activity because of the difficulty of working in a large area open to the weather. In ordinary weather, the surfaces of the concrete were kept wet by water sprayed onto the surfaces whether from a hose held by a workman or by use of water spray pipes attached to the bottom of forms. The spray jets were directed toward the face of the concrete and were spaced to keep the surfaces moist.

In colder weather the contractor CBI was permitted to use 1% calcium chloride in the mix (by weight of cement) to accelerate setting of the cement. At that time, efforts were made to keep the concrete at 60° F. In the fall of 1939 the contractor used steam jets in lieu of water for curing the vertical surfaces and where calcium chloride was used, the curing time was reduced from 21 (low heat cement) to 8 days. Boilers of 1,000 h.p. capacity (total) were used for this service.

As I recall, curing compounds had not yet come into general usage for exposed surfaces of structures in the mid-thirties.

The embedded system of steel pipe with which the concrete was later cooled was placed upon each five-foot layer of concrete with the pipes spaced at about five-foot intervals. The layout was such that a single circuit in the system extended from the inlet (terminating in a gallery or shaft) for a distance of 500 to 600 feet of the coil and terminated in the same gallery or shaft for later connection to the supply and return service water when the cooling was performed. The layout of cooling pipes from any cooling shaft (means of access between galleries) and the spacing of the shafts was such that sections of the dam could be cooled in unison. Water for cooling was taken from the river using several pumps mounted on a barge (moored to the face of the dam) from which the water was pumped into a temporary system of pipes and risers installed in the galleries and shafts. Cooling was started in the fall of 1936 after the west cofferdammed area had been flooded. When the contractor's progress on the reducing width of the dam increased the rate of rise such that the dispersal of setting heat of hydration was retarded, the contractor installed a 5,500 g.p.m. evaporative cooler to enhance the cooling of the concrete. Typically, the cooling operation used about 10,120 g.p.m. and achieved 2,440 tons of refrigeration per day maximum. Cooling water gained about 8° F rise in temperature in passing through the coils. The water was controlled by valves on the terminals of the pipe coils where exposed. Each coil was tested before cooling was started to verify that it did not leak. The coils of pipe of necessity crossed the contraction joints and were joined by use of flexible couplings to accommodate opening of the joints as the cooling progressed. The temperature of the concrete was observed prior to, during, and after the cooling had been completed. Temperature was determined from the water as it exited the coil. Also from thermocouples which had been installed by embedment in the concrete at intervals with the electric leads extended to terminal boxes in the galleries. The cooling commenced as soon after groups of coils could be served then was interrupted until the river temperature dropped so that the final cooling to about 45° F was obtained. Some readings have been taken to determine the long-term experience of the structure. After the cooling was completed and the contraction joints grouted, the cooling coils were filled with grout and the ends of the pipes in the galleries were removed whence the voids remaining were filled by drypacking.

The contraction joints in the dam were filled with grout after the cooling was completed using cement and water mixed to flowable consistency. The mixing plant for the grouting was located under the trestles and piped to points of use. The grout was injected into the feeder end of a system so the grout would rise in the joint beginning at the bottom of the furthermost point from the inlet. Injection continued in each joint or sets of joints until a good flow was returned through the vent pipes connected to the grout groove located across the joints at the level of the gallery floors (50 feet apart). When the joint was filled, the connections were closed and after the grout had set the ends of the pipes exposed in the galleries were removed and the openings drypacked. Thus there is little evidence in the galleries of the dam of the cooling and grouting system. As mentioned in an earlier chapter, the grout was confined to the joint by means of grout stops paralleling the faces, around formed openings and the foundation of the dam. The contraction joint grouting was performed by workmen employed by the government. Warren Simonds, a grouting specialist from the chief engineer's office directed this work. The cement was rescreened through 200 mesh screens at the site to insure a suitable product.

When 260 feet of the right abutment of the dam was removed for construction of the Third Power Plant, it was found (so I am informed) that the contraction joints had been uniformly filled so that the dam had been solidified into a monolithic structure, whereas it had been formed as a series of adjacent great columns of concrete. Photographs seem to confirm the filled joints.

Waste water and waste grout from the contraction joint and deep grouting of the foundation was permitted to flow into the pump sumps in the dam in blocks 31, 53 and 64. From that point the waste water and drainage water from the foundation was discharged from the dam through the 24-inch diameter drain headers extending to the face of the spillway training walls. Later it was discovered that the cement in the waste grout had settled out of the water and had filled the drain, leaving only about six inches of depth at the top of the pipe that remained unplugged. For convenience in removing this hardened grout, it was desirable to get a line through the pipe so that equipment could be used in clearing the excavated material. How to do that? One of the workmen caught the pet cat at the guard shack and tied a cord to its collar while another of the crew went to the far end of the pipe calling, "Kitty, kitty!" But kitty needed a bit of persuasion since the pipe was about 550 feet long, and dark. A jet of compressed air did the trick and the feat was a success. When that story of the cat working at the dam hit the press it generated more interest in the dam than any other story—except that about the flood. Well, the press wanted pictures of the cat. But he was wiser by this time and couldn't be caught. A beautiful, white, silken tabby cat was a stand-in.

CONCRETE — DAMAGE AND REPAIR

After the dam had been put into service, cracks in the concrete were noticed on the face of the spillway. Close inspection disclosed that the cracks were located close to the vertical extremity of the longitudinal contraction joints where such terminated at the intersection with the slope of the dam. Typically, those contraction joints were turned from their normal vertical plane to intersect the downstream face of the dam perpendicularly. That inclined longitudinal joint extended for about three feet in elevation. The grout stop for the joint then lay in that inclined plane to confine the grout in the joint. The crack or fractured areas extended for widths up to 20 to 50 feet in width, as much as 18 inches in depth normal to the dam surface and extended up the face of the dam some 5 to 15 feet. But the cracks did not occur uniformly as to location and I do not know the cause. Nor did I ever undertake to analyze the cause which had many combinations of possible variables including whether the terminating block was to a formed surface or to the adjacent block already constructed, what was the sequence and rate of construction of the adjacent blocks, the temperature of the air and the concrete at time of placement, whether the blocks were subject to flowing water during the diversion of the river, the rate of cooling, the pressure and sequence of the contraction joint grouting, what effect, if any, might be produced by high pressure foundation grouting, whether there was any foundation deformation as the dam came under load from the filling reservoir. I do not have any suggestion as to the basic causes at this writing.

The damaged areas of concrete were removed and shallow areas increased to "square up the void" with minimum depth of refilled areas 18 inches deep by chipping and drilling and barring. Holes were drilled into the underlying concrete and anchors of reinforcing steel installed. The surfaces were carefully sandblasted and after forming,

This view of the east abutment section shows representative concrete activity. At left foreground a concrete "pour" just completed, next right a crew preparing block for the next pour, beyond the connecting catwalk concrete being placed, farther toward center forms raised and secured in place and beyond a workman sandblasting the surface of a block to remove the laitance. Concrete from the mixing plant was dispatched along the trestle in cars moved on tracks by diesel engines ("dinkies") near point of placement. The empty bucket was placed on the vacant space—five spaces, but four buckets to the load—before hooking to the next bucket. Riggers attached and detached the buckets and the loads moved to and from pour by Whirley cranes (no. 3) as directed by the "signal punk." Large "Hammerhead" cranes were also used. Delivery of concrete into the pours directly under the trestle was through the numerous vertical sectional or flexible pipes ("elephant trunks.") April 23, 1937.

113

the concrete was replaced. The repair work was performed in the fall and early spring when the weather was cool to cold. The work was performed by Bureau employees and later by a contractor (Pacific Bridge Co.) The work was achieved from scaffolds attached to the dam with access of men and materials by way of long ladders and skips from cables to hoists on the spillway gate seats. These repaired areas have been exposed to the spillway discharge for nearly 30 years and from present indications, have become permanent fixtures.

The floating caisson for repair of the spillway bucket permitted repair only to about elevation 905 on the downstream face of the dam so an additional caisson called a "face caisson" was used for repair of the concrete surfaces between elevation 905 and 945. The minimum level of the tailwater below the dam was about elevation 935. The repair of the surfaces of the concrete within the face caisson was as described above for other repairs on the downstream face of the dam though the working space was restricted.

EMBEDDED MATERIALS

Reinforcement to prevent cracking of the concrete and to transfer the stress to the adjacent mass of concrete was provided around all openings in the dam. Otherwise the dam was of mass concrete without steel reinforcement. But the powerplants and pumping plants were designed as reinforced concrete buildings and steel as required to carry the stress was installed and held to prescribed location prior to and during embedment by use of auxiliary support frames, anchors or ties. No wood was permitted to be embedded in the concrete such that it could not be removed. Steel pipes for grouting and for foundation drainage, pipe systems for the grouting of the contraction joints, pipes for the cooling of the concrete, metal grout stops extending across contraction joints, anchor bolts for attachment of exposed machinery were also installed in place before the concrete was placed. Such metalwork too, had to be rigidly held so that it could be embedded without damage or displacement. Detailed designs for the reinforcing and all other embedded items were prepared by the chief engineer's staff in advance of need and was furnished to the contractor for fabrication, installation and payment purposes.

This view shows the first elements of the Left Powerhouse sub-structure after the necessary excavation of rock and replacement of concrete in the base structure was to grade. At far left draft tube forms await placement. Next to the right (Bay L3) forms and reinforcing steel are in place. At center the funnels attached to the flexible tubes "elephant trunks" await the start of concrete placing. Next, the foundation of the Station Service Units, and on the far right front is the Control Bay Structure in the making. July 29, 1936.

Construction of Right Powerhouse foundation walls by MWAK makes good progress. June 23, 1937.

CHAPTER VII
CONSTRUCTION

DAM AND POWERPLANT FOUNDATIONS

Early in 1935, the contractor started installing the large steel trestles which would serve the entire foundation area on which concrete would be placed. The upstream trestle with roadway at elevation 1024 would serve the upstream half and the other trestle with roadway at elevation 950, the balance of the work. The footings for the trestle legs were set on and anchored into the prepared bedrock as the excavation advanced. The bents were located so as to interfere minimally with any of the construction features and embedments. The trestle steel legs or bents and bracing was embedded in the concrete of the dam but did not cross contraction joints. The beams and deck of course, lay above the limit of construction so the high trestle remained at the end of the MWAK contract. Both of the trestles were started at the abutments of the dam and extended out toward mid-stream as conditions permitted. The high trestle was completed when the final closure span was placed in the spring of 1937.

Three parallel standard gauge railroad tracks with necessary switching or crossovers for the transport of the concrete and other materials were installed on each trestle. Rails at 28-foot spacing also supported the traveling gantry cranes for delivery of concrete and other materials to the points of use. Utility services—air, water, electricity and telephone systems were mounted under or alongside. Both the hammerhead and the revolving cranes were capable of delivering an 11-ton load at the far limits of the construction. Trackage to serve the concrete shuttle cars was extended to the mixing plants and beyond. Lumber, prefabricated forms, reinforcing steel and all other materials going into the construction was delivered by truck within reach of the cranes for placement where needed in the work. The contractor MWAK, also extended the construction railroad onto the high trestle to deliver certain rail shipments directly to the dam and a temporary railroad spur line was constructed along the forebay slope to carry concrete to Block 40.

Concrete to serve the downstream trestle was delivered from the mixing plants by inclined skipways for transfer to the lower levels. Those trestles were removed and replaced with a single trestle at elevation 1180 by CBI. From that one trestle, the entire width of the dam was served.

When the abutment portions of the dam had been built to elevation 1220 a seat was formed in the concrete to support the upper rail and wheels of the gantry for the whirley cranes, the lower wheels riding the 1180 trestle upstream rail. This "side-hill" gantry mount was necessary to reach the trashrack structures.

In my discussion of the details of the various construction subjects I have not endeavored to specifically separate the work between the various contractors who performed the work except incidentally to indicate the "state of the art" at a particular time. The result is that discussion of some subjects cover the entire period of construction.

THE HIGH DAM — CONTRACT MATTERS

By the time the work was nearing completion on the MWAK contract, designs and specifications were ready and specifications (No. 757) issued for completion of the High Dam. Bids were opened December 10, 1937 with the joint venture sponsored by the H.J. Kaiser Co., the low bidder at $34,442,240. The other bidder, Pacific Constructors, Inc., bid almost 23% higher. The low bidder included the MWAK Co. and the Six Companies (which had constructed Boulder/Hoover Dam.) Those latter joint venture firms were Morrison-Knudson Co., Henry J. Kaiser Co., Utah Construction Co., J.F. Shea Co., McDonald and Kahn Co., and Pacific Bridge Co.

MWAK still had not finished a few items of its contract. Also MWAK had a number of unsettled claims that it believed were due. The contract for completion of the work was awarded to the low bidder then known as the Consolidated Builders, Inc. on February 7, 1938.

The Bureau had exercised its option for the MWAK camp, rolling stock, plant and equipment and then terminated the MWAK contract on March 21, 1938. On that same date CBI (as it was abbreviated) received and acknowledged Notice to Proceed with the work to be completed by March 20, 1942. CBI agreed to complete the unfinished items of MWAK work for $48,000 under change order #1, which amount was deducted also under a change order from MWAK

Spokane Chamber of Commerce hosts the successful bidders Specifications No. 757.

Seated:

Les Corey, Vice President, General Manager Utah Construction Co.

Henry J. Kaiser, President, The Kaiser Co. and Six Cos.

Tom Walsh Sr., President Walsh Construction Co. and MWAK Co.

Guy F. Atkinson, President, Guy F. Atkinson Co. and Executive Vice-president of the MWAK Co.

Charles A. Shea, President (?) of J.F. Shea Co.

Francis Donaldson, Vice-president of MWAK Co.

W.E. Kier, Vice-president and Secretary of MWAK Co.

Standing:

Leo J. Fischer, Pres. Thompson-Starrett, Director MWAK Co.

Ben H. Kizer, Attorney of Spokane

Phillip Hart, Pres. Pacific Bridge Co.

Edgar Kaiser, later General Manager of Consolidated Builders, Inc.

Tom Price, of the Kaiser Co.

Harry Morrison, Vice-president and General Manager, Morrison-Knudsen Co.

Felix Kahn, of MacDonald & Kahn

David Small, Vice-president, Walsh Const. Co., Director MWAK Co.

J.A. (Jack) McEachern, General Construction Co.

Paul Wattis, Utah Construction Co.

Col. M.J. Whitson, Director and vice-president MWAK Co.

George H. Atkinson, General manager, MWAK Co.

Gil Shea, Vice-president, J.F. Shea Co.

Arthur G. Moulton, Vice-president Thompson-Starrett and Vice-president, MWAK Co.

Photo by Libby—permission Cheney Cowles Museum, Spokane.

final payment. Final payment total earnings (net) $38,172,560. MWAK filed suit against the United States Government in the U.S. Court of Claims for $5,267,273 seeking additional compensation for work asserted to be required in excess of the contract provisions. Hearings before the Court of Claims Commissioner were held in Mason City, Washington, D.C. and Flagstaff, Arizona. The suit was finally dismissed by the court on January 8, 1947. (I was detailed to the Department of Justice in Washington, D.C. to assist in this case in preparation for and during the hearings.)

Odd reactions often come from too much concentration. When we were in Flagstaff John Kier suggested that we might go out to the Navajo Indian Reservation Naval Ammunition Depot on our "Sunday off." Whereupon Commissioner Ramseyer queried, "What do the Indians want with all that ammunition?"

The work required to be performed by CBI included the care and diversion of the river during construction; completion of the dam to its full height; excavating (by change order) the tunnels for the future discharge pipes for the irrigation supply; excavating for and constructing the wing dam at an angle to the main dam and embed therein 12 intake tubes for the future pumping units. (The wing dam to serve as the foundation for the future pumping plant); building the Left Power Plant structure and place the concrete about the turbines and generator machinery (installed by others); build the Left Switchyards; and other related features; embed the penstocks installed by others. The Bureau retained 59 houses from the camp in Mason City when the camp—houses, shops, store buildings, mess hall, recreation hall, restaurant, garage, motel—rolling stock, aggregate facilities, mixing plants, cement storage, conveyor systems, utilities, etc., (which it acquired from MWAK) was turned over to CBI upon termination of MWAK's contract.

After award of the contract for completion of the dam and appurtenances, CBI completed the items of "cleanup" remaining from the MWAK job on the Change Order finishing the removal of the cofferdam and the cribs from Blocks 39 and 40 and accumulated debris from the spillway bucket, etc. Also, the contractor rebuilt and replaced the mixing plants and made extensive repairs to the aggregate facilities which had been apparently purposely neglected by MWAK as its work neared completion. The time was well used awaiting the construction of the high trestle by the Bethlehem Steel Co., after

which concrete construction under CBI's contract could be resumed.

After completion of the Spillway bridge the contractor removed its trestles, placed concrete in the formed shallow openings where the trestle legs had been cut off, and after completing the Right Power House on an Extra Work Order, removed his plant and equipment and turned the camp back to the Bureau of Reclamation.

CBI completed all of its work and the job was accepted January 31, 1943 with total contract payments of $41,366,900.

THE "DAM" COMPLETED

When the "force account" work was terminated in 1947 the Bureau moved promptly and the Chief Engineer prepared specifications and called for bids for several parcels of work needed to complete the dam and the appurtenances contemplated in the existing concept for the site. Specifications No. 2075 for the completion of the transformer circuits, the tie circuits and the right switchyard and other items were issued and bids were received February 25, 1948. A joint venture of Morrison—Knudsen and Peter Kiewit companies was the successful bidder for $2,789,284. And as that work was getting underway, Specifications No. 2329 was issued and bids received for the completion of the Pumping Plant, Bus Runway, Raising the Spillway Training Walls, Excavation for the Feeder Canal, Construction of the Water Supply Reservoir; Pump-Discharge Outlet Structure; Installation of Pump Discharge Pipes and other items. Again, M-K/P-K was the successful bidder at $13,348,449. Both jobs were managed by a single organization with Ole Strandberg as the contractor's general manager. The government had made its cement handling and storage facilities, aggregate and concrete mixing plants and certain other plant and equipment available to the contractor so time was not lost in getting the job moving. This contractor also operated the 30-mile railroad to the dam and furnished concrete to the government and others. The placing of second stage concrete about the turbines and generators in the Right Power Plant was included by change notice before the bids were received. These contracts were seemingly more difficult to administer because of the short time available to the Chief Engineer to do the designs and specifications before issuance. Also the work was widely scattered over the site and involved hundreds of items of work for payment purposes. There were numerous "tests of will" on

Sta.4+74 = Sta. PIO+00
Intersection of axes of dams
Control cable gallery
Bus tunnel El.1259.5
Elevator tower
Top of dam-El.1311.08
Gate hoist transfer structures
Elevator tower
135'x

senger elevator
ft-Sta. 2+59
El.1240.0
El.1171.0
40

£ Drain from pumping plant
£ Hoist gallery
£ Control cable gallery El.1252.5

dam

Excavated rock profile

Station service trashracks
Main unit trashracks
S=0.8:1
10
20
30

SECTION-STA. 5+23
BLOCK 7

UPSTREAM

£ Stairway to galleries Sta.43+46
Elevator tower
Top of dam El.1311.08
Parapet El.1315.0
Elevator tower

El.1273.0

90
80
70
Sta. 30+70
60
50
Spi

UNDEVELOPED POWER PLANT

0   100   500
SCALE OF FEET

DOWNSTREAM

Top of dam El.1311.08
£ Hoist gallery
Crane rails

£ Longitudinal inspection galleries

S=0.8:1

Control cable gallery

Developed power plant

6'-0"Dia.penstock

Station service units
El.938.0

12,500 Kv-a Generator
14,000 H.P. Turbine

£ Grouting and drainage gallery

SECTION-STA. 7+63
BLOCK 12

El.1290.0
Top of dam El.1311.08
Crane rails

Main unit trashrack

S=0.8:1

Control cable gallery

18'-0"Dia.penstock

El.1026.0
El.1015.0

Developed power plant

120,000 kv-a Generator

El.938.0
147,000 H.P. Turbine

S=0.15:1

Axis

SECTION-STA. 8+44
BLOCK 14

Elevator tower

Max.W.S. El.1290.0
El.1293.0
Crest El.1260.0
Elevator shaft
Spillway training wall
Transverse galleries

Control cable gallery

Developed power plant

El.1034.0

S=0.8:1

S=0.15:1

Axis

SECTION-STA. 14+20
BLOCK 31

Max.W.S. El.1290.0
To
Crest El.1260.0

El.1160.07
El.1136.67
Outlet work trashrack

El.1036.67

El.980.0

El.935.76
El.900.0
El.910.0
El.869.68

S=0.15:1

Axis

SEC

118

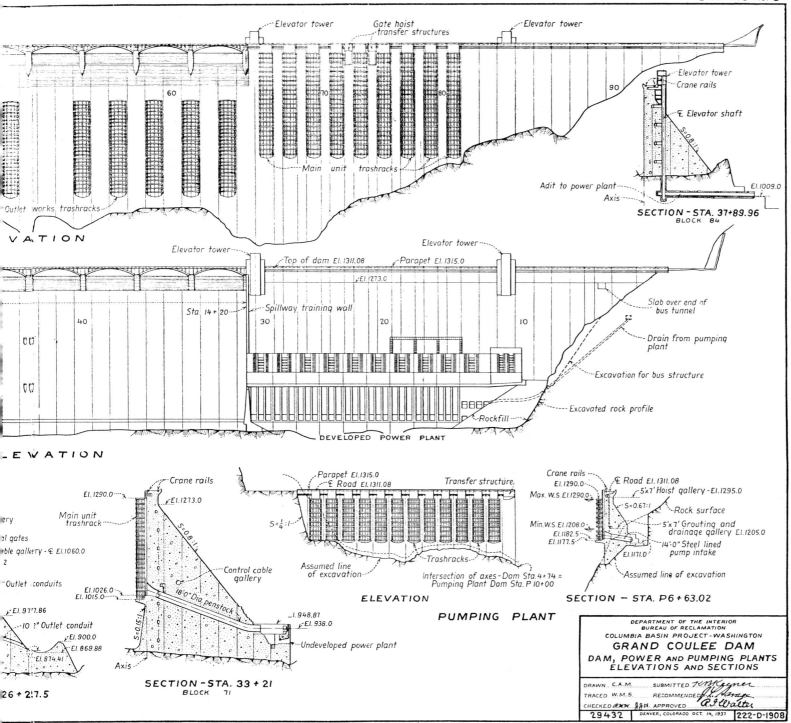

Elevator tower
Gate hoist
transfer structures
Elevator tower

60   70   80   90

Main unit trashracks

Outlet works, trashracks

VATION

Elevator tower
Crane rails

£ Elevator shaft

Adit to power plant
Axis

El.1009.0

SECTION - STA. 37+89.96
BLOCK 84

Elevator tower
Top of dam El.1311.08
Parapet El.1315.0
Elevator tower

El.1273.0

Sta. 14+20
Spillway training wall

40   30   20   10

Slab over end of
bus tunnel

Drain from pumping
plant

Excavation for bus structure

Excavated rock profile

Rockfill

DEVELOPED POWER PLANT

EVATION

Crane rails

El. 1290.0
El.1273.0

Main unit
trashrack

S=0.8:1

Control cable
gallery

18'0"Dia. penstock

S=0.15:1

18'0"Dia. penstock

I.948.81
El. 938.0

Undeveloped power plant

El.1026.0
El.1015.0

El.977.86

10 ?" Outlet conduit

El. 900.0
El. 869.88

El. 874.41

Axis

SECTION - STA. 33 +21
BLOCK 71

Parapet El.1315.0
£ Road El.1311.08

Transfer structure

S=1/4:1

Assumed line
of excavation

Trashracks

Intersection of axes-Dam Sta.4+74 =
Pumping Plant Dam Sta. P 10+00

ELEVATION

PUMPING PLANT

Crane rails
El. 1290.0
Max. W.S. El.1290.0

Min.W.S. El.1208.0
El.1182.5
El.1177.5

£ Road El.1311.08
5'x 7' Hoist gallery -El.1295.0

S=0.67:1

Rock surface

5'x 7' Grouting and
drainage gallery El.1205.0

14'-0" Steel lined
pump intake

El.1171.0

Assumed line of excavation

SECTION - STA. P6 + 63.02

ery

l gates

ble gallery - £ El.1060.0

Outlet conduits

26 + 2!7.5

DEPARTMENT OF THE INTERIOR
BUREAU OF RECLAMATION
COLUMBIA BASIN PROJECT-WASHINGTON
GRAND COULEE DAM
DAM, POWER AND PUMPING PLANTS
ELEVATIONS AND SECTIONS

DRAWN C.A.M.   SUBMITTED
TRACED W.M.S.   RECOMMENDED
CHECKED   APPROVED
29432   DENVER, COLORADO OCT. 14, 1937   222-D-1908

these contracts and it seemed to me that much of my time and that of the contractors was spent in pursuing minute items of difference and claims for additional compensation. When severely cold weather set in January 1949, the contractor decided to shut down concrete activity on urgently needed items, taking the position that payment of liquidated damages for the delay was less costly than continuing to fight the freezing weather.

The contractor was directed to proceed in accordance with the approved schedule and filed a protest and claim for additional compensation. I do not know the outcome.

After construction of the transformer deck trackage and cranes were placed thereon to service the construction of the Left Powerhouse. Concrete was delivered from the high trestle via the inclined skip transfer cars. Note concrete "hand" riding the bucket and dumping concrete through the "elephant trunk" into the forms. May 29, 1939.

The removal of the cofferdam downstream of blocks 39 to 41 seemed to be more of a task than planned for. February 22, 1939.

CHAPTER VIII
SPECIAL FEATURES

TWIST SLOTS

Design studies using "twist structure analysis" indicated that the monolithic structure contemplated to be achieved after cooling and grouting the contraction joints of the dam would respond to the forces introduced by the filling of the reservoir causing a concentration of stress in the portion of the dam at the sharply rising abutments. It was desirable to relieve such stresses to reduce the risk of cracking the concrete in the initial filling of the reservoir.

Five twist slots, as they were called, were constructed in the dam abutment sections to permit the structure to adjust to its load while it had some flexibility. Twist slots to be filled with sand were constructed in Blocks numbered 7, 9, 78, 82 and 87.

The concrete was formed to leave openings six feet wide except at the faces of the dam and around gallery openings where cantilever construction provided continuity, and at 50-foot intervals, vertically. The slot opening was narrowed at the 50-foot levels to 12 inches. As the concrete in the adjoining blocks advanced in height, the twist slots were filled with sand, consolidated to maximum density. When the filling reached gallery level, (50 foot intervals), the narrow gaps were spanned with heavy timber supported by bolts from steel channel sections resting on the cantilevered surface which carried the sand filling.

After the reservoir had been filled to elevation 1150, the sand in the twist slots was removed and the slots filled with concrete. The removal and filling was performed taking the sand from the lowest 50 feet first, and it was refilled with concrete prior to removal of the sand in the portion above. That sequence was used to retain lateral support across the slots. Access to the slots was provided by deferring construction of the floors of the galleries until the slots were complete at each level.

SPILLWAY TRAINING WALLS

The water flowing over the spillway of the dam is confined from the adjacent power plants transformer decks by the high training walls at each end of the spillway. The walls were designed using the best data then at hand—but there had never been a spillway with the discharge or height and depth of this one. It soon became apparent after the spillway came into use that the walls needed to be raised because the spray from the "boiling bucket" and from the entrainment of air as the water dropped down the face of the dam caused a constant rain to fall behind the power plants. The walls were raised twice and then came the great 1948 flood, after which the training walls were raised again. At the outset, no one would have imagined the walls of their present height.

The spillway bridge provided the final means for crossing the Columbia River at Grand Coulee Dam. Note the east training wall still under construction at left. November 22, 1941.

The spillway bridge arches now complete and construction of spandrel walls and roadway under way. Construction of remainder of dam was nearly complete. August 23, 1941.

SPILLWAY BRIDGE

The spillway bridge was constructed on timber falsework resting upon the drumgates which were in the lowered position. The work on the bridge was serviced from the trestle using the large revolving cranes. After forming and placement of the reinforcing steel, the arch sections of the bridge were cast in sections, leaving unfilled gaps to reduce the shrinkage effects on the arch length and to avoid stress and settlement problems as the increased weight of the arch being built compressed the supporting elements of the falsework. After restoring the forms to proper elevation, the gaps were then filled progressively. The last segments to be concreted in each arch were those farthest from the piers. After the arch sections had attained sufficient strength, the supporting walls and roadway were constructed completing the link in a new passageway across the Columbia River. The bridge was designed to carry the maximum loadings anticipated when transporting transformers, turbine and generator parts for the 108,000 kw units for the Right Power Plant. That design was based upon concrete having minimum compressive strength of 3000 pounds per square inch at age 28 days.

When planning the construction of the Third Power Plant at the dam, the transport of the larger parts across the river required more than passing attention. The anticipated maximum loads were 282 tons as compared to 90 tons for the previous plant. Several schemes were considered but at the suggestion of consulting architect, Kenneth Brooks, (who was also employed to prepare a master plan for the project) to consider strengthening the bridge arches, cores were taken from the bridge concrete. The then (1966) compressive strength of the concrete in the roadway was found to be 8160 psi (pounds per square inch), and 7630 psi in the walls. In 1941 the 28-day strength of the concrete was 5140 psi.

The high quality of the concrete permitted use of the bridge as it was and the loads were transported slowly on special trailers that reduced the wheel and axle loads so that the structure was undamaged—and at substantial savings.

The truck and trailer used for such transport had 90 tires, a wheel base of 101 feet and a width of over 17 feet.

PUMPING PLANT

The plan for the irrigation of Columbia Basin Project lands contemplated a pumping plant of 12 units of 1600 cfs each powered by bus connected motors of 65,000 hp each. Six pumps were installed initially and by the time additional pumping capacity was deemed necessary, the value of peaking power had increased sufficiently to warrant installation of pump/generator units so that those units could

The spillway bridge spans were supported by timber framework resting upon the completed drumgates. July 31, 1941.

The future Pumping Plant site was this pit behind the wing dam constructed by CBI. The pump inlet pipes were closed by hemispherical bulkheads. Discharge pipe tunnels, excavated through the back wall of the pit. The backwall was later given additional support against displacement by installing large, steel dowel members in holes drilled for that purpose. July 23, 1940.

be operated off peak as pumps and as generators during period of peak capacity demands. A change in design followed and six pump generators were added, each of 50,000 kw capacity, thus adding 300,000 to the capacity of the power installations at the dam. However two of these units, of foreign manufacture, have not lived up to expectations. While operating as generators, water flows back up the feeder canal from Banks Lake with the result that Banks Lake is maintained at higher average levels than before. Also, since the advent of the Canadian storage above the dam, the power used for pumping the irrigation supply no longer uses primarily secondary power of the flood season. Originally, the feeder canal was constructed using twin barrel cut and cover design through ground which appeared to be too unstable to warrant the normal trapezoidal section of canal through that reach. When the turbine/generators were installed the feeder canal was rebuilt to more readily accommodate the needs of reverse flow in the canal and the cut and cover section was rebuilt as an open waterway, bench flume type.

Base of Training wall at end of Left Powerhouse—forms in place for continuation. Note the curved section of the spillway bucket being formed into the base of the training wall. Workmen removing loose rock and remaining earth cover from foundation rock prior to drilling and blasting. June 9, 1936.

LIFT SEAM DOWELLING

When the excavation of the pumping plant was being performed, a so-called "lift-seam" in the rock was noted. This extensive sloping joint structure in the bedrock in the back wall of the pumping plant was viewed as a potential hazard in the event of an earthquake. To alleviate this risk 46 six-inch diameter holes were drilled in the rock into which steel dowels were inserted and grouted into place. The dowels were composed of eight 1¼-inch and two 1-inch steel reinforcing bars welded together. "Rock bolting" had not yet become a common "tool of the trade."

PAINTING

All exposed ferrous metal work was either painted or had been galvanized prior to delivery. Typically, Bureau practice had been to use coal tar products for protection of metal subject to contact with water. In earlier years, hot applied coatings had been used when feasible because of more favorable results but in the 1930s the chief engineer's research personnel began the development—in cooperation with some paint manufacturers—of "better" coal tar coatings that could be applied at normal air temperatures. Also other coating processes began to appear in the marketplace that held promise of good protection, relative ease of application (though critical control) and reasonable cost. Products of various types were used in tests on the metal surfaces at the dam with mixed results. Typically though the metal surfaces exposed to water were coated with tar products. Trash racks were hot dipped at the storage site and placed in the slots in the rack structures later. The exposed surfaces of metalwork partially embedded in the structures were painted with a coal tar paint developed in the chief engineer's laboratory and identified as CA-5. Prior to painting the surfaces were carefully cleaned of all loose scale, rust, oil, etc. The cleaning was accomplished by wire brushing, sand blasting and flame cleaning. The latter method was used on the drum gate surfaces. Gas fueled torches with multiple flame ports were effective on large, flat surfaces in removing millscale from the steel. Where the surfaces were exposed to high velocity flows such as in the penstocks and outlet tubes, the surfaces were grit blasted. The downstream faces and the pier plates of the drum gates were sandblasted and coated with a more sophisticated coating with the prospect of better service where it would be in pressured rubbing contact with the gate seals.

Aerial view looking down into the discharge pipe tunnels for the pumping plant. Cranes at center of view were used in construction of the pumping plant building in the pit behind the wing dam and adjacent the canyon wall. June 15, 1949.

One of the 102-inch cast steel outlet liners enroute to its location as the gate assembly proceeds. Cast steel linings extended from the gates to the upstream face of the dam and downstream to the longitudinal contraction joint of the block. Beyond that joint the liners were of plate steel. July 20, 1939.

CHAPTER IX
OPERATING FEATURES

OUTLET CONTROL WORKS

To permit the discharge of water through the dam during construction and later during reservoir operations, 60 discharge tubes, 102 inches in diameter were built into the dam. The control for such was by use of slide or tracked gates in tandem for each outlet tube. These high pressure gates consisted of cast steel bodies extending above and below the tube, the upper and lower bonnets and the moveable gate section below which was a ring or open section, and cast steel tubes extending to the face of the dam and downstream some 15 feet from the gate. Hoisting mechanisms were a part of the bonnet and drains were extended from the bottom of the gate body to the gallery below to permit dewatering. Hoisting mechanisms were located in the galleries at 950, 1050, and 1150.

The designs for the paradox gates for the lower set of outlets in the dam, being new concepts, took more time in design than normal and so the bid call for the gates was late. Hardie-Tynes Manufacturing Co. of Birmingham, Alabama was the successful bidder. And the time for delivery was short. The gates were required for the successful prosecution of the work at the dam. For the gates would be used for passing a portion of the river flow as soon as installed. Mr. Savage was so interested in the timely delivery of the gates that he sought and received the personal assurances of the president of the Hardie-Tynes Co. that the gates would be delivered promptly—but they were late. The portions of the outlet tubes downstream from the gates in the upper two tiers were lined with plate steel since those were to be for normal usage. The tubes for the lower outlets were formed into the concrete and not lined since they were to be used only rarely after the dam was in service. The surface of the concrete of this set of tubes, while well formed, did not have a surface condition suitable for exposure to the high velocity streams of water that would follow. Temporary bulkheads were placed in the downstream ends of the outlet tubes before the cofferdammed area was flooded. This

permitted access in the dry from the 950 gallery. The surfaces of the tubes and the contraction joints therein were then finished after the contraction joints adjacent were grouted. The resurfacing of the concrete consisted of sandblasting the same to remove the mortar coating and expose the bare, clean sand and coarse aggregate. A mortar coating was then applied and, using hand-held carborundum stones, the surface was thoroughly rubbed until the surface was dense, hard and smooth.

The lower set of 102-inch outlets were formed into the concrete and not lined with steel beyond the gate elements. August 26, 1937.

The downstream ends of the middle and upper sets of 102-inch outlets were curved downward toward the spillway bucket. Working near the rushing water was a mental hazard endured by all. June 12, 1939.

Drum gate chamber formed in the concrete. The gates were hinged on the right and when in the lowered position the gate rested on the seat at left (downstream). Pier plates embedded in the concrete are heated by electrical resistance heating elements during freezing weather to avoid damage to seals. August 27, 1940.

The work was performed in the dead of winter and, of course, involved hard, manual work in an enclosed, damp area. At the end of shift the tired and perspiring workmen climbed out of the dam by ladders and stairways to the cold air—many developed colds and they complained of the hazards of the environment of the work place. In discussions with their steward, I learned that they had decided that their trouble was caused by the humidity—so I agreed to measure it. With psychrometer in hand, I determined it in their presence and when I stated that the humidity there was 100%, the steward asserted, "Well, you surely can't beat that." The work dispute was ended.

During gate operations, while the gate was moving from closed to open, extreme turbulence of the water in the gate body occurs and any foreign materials being carried in the stream can become deposited in and around the gate parts. Before flooding, the areas of the inlet and the trashrack structures were thoroughly cleaned of all loose and foreign material. But as work continued on the portions rising above the water, debris falling from the activity passed down into and around the trash racks and inlets causing severe damage to the gate mechanism during

opening and closing. The gates were equipped with roller trains which greatly reduced the frictional forces when the gate was operated but with bolts and nuts, nails, wire and other metallic fragments and sand and gravel coming between the rollers and the smooth tracks on which they moved the tracks and rollers became severely pitted and in some cases the movement of the tracks was prevented when displacement occurred and the rollers fouled up on the gate gussets causing the rollers to slide.

Extensive damage to these parts required modifications to the design and extensive repair to the damaged parts.

Another unforeseen development occurred when the upstream gates of the lowest set of outlets were closed and the downstream gates were not subject to the pressure of the water in the reservoir but were subject to the surging of the tailwater when the spillway was discharging the flood flows. The continuous repetition of stress on the bolts caused them to fail by metal fatigue. After repair, these downstream gates then were left in the closed position.

When the Third Power Plant was built, the lower sets of gates were removed and the outlet tubes were plugged. The 950 gallery was then used for power cables to the switch yard.

The rust and the corrosion of metal parts exposed to water is a curse to be avoided, if possible. Since satisfactory coatings for metal parts were hoped for, test coatings were applied in one of the steel linings of the outlet works. The tests made in cooperation with manufacturers were intended to indicate the most economical, serviceable, cold-applied coatings available for exposure to high velocity water discharges as well as to the moisture and condensation present when discharge was suspended. Epoxy products were not then available.

A disaster of potentially major proportions occured on March 14, 1952 when one of the outlet gates was opened while the manhole giving access to the outlet was unsecured. The operator of the gate panicked when a 30-inch stream of water came shooting up into the gallery and instead of reclosing the gate, made a quick retreat. In short order, the galleries in the Dam were flooded up to the 950 level and water started pouring into the power houses, also flooding them to levels above the turbines. Oil floated out of the bearings and the entire plant was in danger. If the flow had not been stopped, the water would have run out the power house doors at elevation 1012, well above the generators. Ultimate disaster was prevented by three courageous engineers, Norman Holmdahl, Don McGregor and Milton Berg. When aware of what was happening, they hurried to the top of the dam, entered, rushed down the stairs to the 1050 gallery and waded through a rushing stream of cold water toward its source, only to find that they were on the wrong side of the huge water jet—30 inches in diameter—thrusting up into the gallery. Retracing their steps, they raced up to the gallery above and crossed through the dam to the next stairway down to the 1050 gallery then again forced their way against the stream—now waist deep—until they came to the gates in Block 55, whereupon they closed the switch which was then under water and opened the downstream gate to relieve the pressure on the column of water entering the dam. They then moved to the upstream gate and closed the switch to cause that gate to reclose. The gate control motors and mechanism, though under water, reponded and the disaster was controlled. Thereafter, closures were placed in the galleries entering the power houses from the dam to prevent a future flooding of that kind or from a failure of any of the features of the dam. This episode was the subject of an article in a 1952 issue of *Readers' Digest*.

CAVITATION IN OUTLET WORKS

After the 102-inch outlets were used for diversion of the river, inspection of the unlined transition from the steel lining to the face of the spillway disclosed erosion of the concrete surfaces by "cavitation." Such erosion occurs from forces developed when flowing water at high velocity is diverted from its trajectory. The phenomena can usually be relieved and the forces reduced by the addition of air to the zone of contact where the forces and the damage are caused or concentrated. To relieve this condition in the outlet works, a "collar" was excavated in the concrete at the downstream end of the steel lining and air was supplied to the collar by piping then embedded in trenches excavated in the face of the dam. Openings in the piping to the face of the spillway admitted air to the collar and to the discharging stream.

DRUM GATES

The flow of water over the spillway is controlled by eleven drum gates, each 135 feet in length, with a rise of 28 feet. Removable flashboards increase the controlled height to 30 feet, attaining the maximum reservoir height. The gates were constructed of steel plates fabricated in the bays of the spillway. The gates are large, hollow, water-tight structures with two of the sides arc-shaped, the other being flat. They are hinged at the upstream edge and are floated in a chamber of water. By use of differential level control valves in the water from the reservoir and with delivery and discharge lines connecting the chamber, the gates can be raised or lowered or held in any partially open position at all levels of the reservoir above elevation 1255. The gates are attached to the concrete of the dam by large hinges of cast steel anchored by large, long bolts extending down into the concrete. Steel pins extending through the hinge bearings permit the movement of the gates. The gates initially were controlled in shafts within the piers of each gate. Later, this control was revised so that the gates are operated remotely from the control room in the Left Power House. Flexible sealing members of suitable rigidity are installed along the edges of the gates and gate chamber to retain the water within the gate

As the flow into the reservoir declined after the summer flood, the entire flow was again passed through the outlets. Work on the drum gates began. Meanwhile, some of the abutment blocks had been constructed to the roadway level. Note the "side-hill" gantry crane for servicing the top of the dam and the trash rack structures. October 2, 1940.

Drum gates number 6 and 7 in the lowered position. Reservoir rising in the spring flood. May 9, 1941.

Drum gate assembly nearing completion. April 9, 1941.

Drum gate on the right being painted. Gate in center has been painted and hydraulic hoisting devices in place for lowering the gate into the chamber. May 6, 1941.

Ironworkers assembling drum gate. Pins in rivet holes maintained alignment and position of sections of the gates until bolting and then riveting could begin. Reinforcing steel protruding—upper center—is for spillyway bridge arch where it abutted the pier. December 7, 1940.

chamber. These seals are in contact with the painted surfaces and are a source of abrasion requiring occasional repainting. The gates were assembled with the flat section within the gate chamber and with the hinge members in proper horizontal position supported on timber falsework erected within the gate chamber. The gate elements were first bolted tightly together, the gate was then checked for position and the bolts removed progressively as the connections were riveted to complete the structure. When complete, large hydraulic hoists were attached to the downstream surface of the gate, permitting the removal of the timber falsework and the debris from the gate chamber. The gates were next lowered into the water-filled chamber and with the gates thus supported, the temporary hoists and attachments to the gates were removed.

The completion of the drum gates before the spring flood overtopped the gates was a matter of much concern. And when there seemed to be "too many rivets requiring replacement," Edgar Kaiser came down to the job and discussed the matter with Bill Morgan, the Bureau inspector in the presence of the iron worker bosses, Claude Bacon and Scotty Wright. Mr. Morgan explained that loose rivets were not acceptable. Mr. Kaiser, as he was leaving, said, "Well, Mr. Morgan, you will admit that Claude Bacon is a pretty good man!"

"I am sorry, Mr. Kaiser, I only inspect the work!" was the reply.

Vexing problems with the drum gates have been the occasional failure and leakage of the seals. The seals are attached to the downstream seat and for those on the ends of the gate are attached to the gate. The seals prevent the chamber from becoming underwatered when the gates are in raised position. The seals are of durable rubber and are faced with spring steel shaped to maintain sliding contact with the gate and pier surfaces at any gate position. Originally, the downstream faces of the drum gates were to be coated with waterproof grease. Now—1986—those surfaces of the drum gates are coated with "Elastuff" which is giving good service after seven years of exposure.

The drum gates as originally built, when in the fully-raised position, had a lip elevation of 1288, 28 feet above the crest of the spillway. Since the maximum level of the reservoir was at 1290 feet above sea level, decision was made to add flash boards onto the lips of the gates so that the controlled level of the reservoir was then 1290. This action permitted increasing the reservoir capacity for storage by 163,200 acre feet and increased the controlled head at the site by two feet, which had a significant increase in the output of power from the powerplants. After the Canadian Treaty storage dams were built, they took 15,500,000 acre feet of water off the annual flood flow in the Columbia. And after the Third Power Plant began operation with its greatly increased hydraulic capacity substantially all of the flow of the river is now passed through the penstocks (and outlets when storage water is released for flood control.) Also the flash boards are not removed at the beginning of each flood season as in the past. To avoid any problems with Canada, care is taken so that the reservoir is held at elevation 1290 or below.

I recall being awakened at 2:30 one Sunday morning when the operator called to advise me that the reservoir had risen to elevation

1290.3. I told him what gate adjustments to make and he replied that he had already done so—that he had gotten his instructions from George Owens. When I asked why he had called me, he said that he had called to apologize. He had reread the instructions and he was supposed to call me first. Thanks a lot!

PENSTOCKS

The penstocks for the power plants were of plate steel rolled or shaped to fit at the rolling mills of the supplier and shipped to the job in sections or pieces as large as could be economically shipped by rail. The fabricator and erection contractor then completed necessary welding and assembly of sections of pipe at an assembly yard along the railroad near the present town of Electric City. All welded seams or joints were then viewed by means of X-ray techniques with individual coded markings showing on the X-ray pictures. Welding techniques and results had improved substantially in the mid-1930s but subsurface and surface imperfections did occur. When noted in the field or by careful review of the X-ray pictures, the faulty areas were chipped out and rewelded and X-rayed again. Repairs delayed the work, of course, and some of the fabricating contractor's personnel took short cuts by substituting pictures of good weldments with codes on the film for other parts of the joints so that defective welds would not appear in the film submitted to the Bureau personnel. The work of assembly had advanced substantially before one of the engineers or technicians recognized that some of the pictures, while coded differently, were actually of the same weld. Since the welds are as distinctive and identifiable as fingerprints, the contractor's fraudulent scheme came to light and the entire assemblage had to be re-X-rayed and defective welds repaired. This deceptive practice necessitated far more control of the verification for acceptability of the work. I do not recall how the guilty parties were brought to account. The penstock inspection had not been assigned to Bert Hall or his team.

Many years later, when work of building the Alaskan Pipeline from the Arctic-Prudhoe Bay oilfields was in progress, I read of the discovery of a similar fraudulent episode by contractors there engaged.

The assembled sections were transported to the dam and—other than the upstream section—were placed into the tunnel-like openings

130

that had been formed in the concrete during the MWAK contract. At the inlet, the first section of the penstock was closed by a temporary hemispherical steel bulkhead welded into the end of the steel pipe. This was set some distance from the face of the dam and the connection from the reservoir was a smooth transition section or inlet formed in the concrete. The first section of each penstock was embedded in the mass concrete. The remaining sections of the penstocks were then brought into the tunnel from the downstream end and when secured and welds complete and verified, the openings about the penstocks were filled with concrete delivered into the annular space about the 18-foot diameter pipes, using "Pump Crete" machines wherewith the concrete was pumped to points of use. Such machines had been available to the construction industry for a number of years. Before putting into service, the penstocks' interior surfaces were blasted with steel shot or grit and painted. The lower end of the penstock was welded to an expansion

Reinforcing steel cages enclosed the forms for the 24-foot diameter octagonal tunnels or openings in which the penstocks were later placed. September 21, 1938.

The upstream section of each 18-foot diameter penstock was temporarily closed by the hemispherical bulkhead. This section was embedded in the mass concrete while the remaining sections of the penstocks were installed in the octagonal openings, the forms of which can be seen taking shape within the reinforcing steel hoops at the right. The water passage from the steel penstock to the face of the dam was formed into the concrete. October 24, 1938.

joint bolted to the inlet of the scroll case for the turbine. The inlet of the penstock is protected by means of trash racks extending to the reservoir level. The water entering the penstock can be closed off using large, hydraulically operated gates over the inlet. Present too, are pipes and valves whereby after the penstock has been emptied of water, it can be refilled with water from the reservoir and the air can be exhausted by an air vent, extending above the maximum reservoir level.

PENSTOCK STOPLOGS

When attempt was made to position the stoplogs for one of the penstocks to repair leakage about the embedded gate seal, the bottom log would not descend to the bottom of the seat at elevation 1015. Our contract diver, Colin O'Donnell, made an unprecedented descent of 275 feet to discern the problem and detailed it when he returned to the surface. Communication with a diver in 275 feet of water seems to affect the wave length of the sounds from the "voice box" and he could not be understood while at such depths. Also the working time of a diver at such depths is only a few minutes per day and is a dangerous occupation. The diver reported an accumulation of debris from the construction of the trashrack structures was lodged on the stoplog seat, which necessarily must be removed. Devices were fabricated to control a high pressure, submersible pump with a series of jets to remove the accumulation Then a template was constructed with many protruding "fingers" which, when lowered onto the seat, would retract under the pressure of any remaining obstructions. When brought to the surface, the locations were noted and the pump again lowered until the seat was clean. Now I suspect that such inspection would be made with a television camera.

POWER PLANTS

The foundations for the power plants were constructed by MWAK with the work carried to an elevation—948.87—sufficient to permit extension of the buildings without cofferdams. The excavation and concrete construction was conducted concurrently with the building of the dam on which the power plants would rest. Each of the plants included a service bay, an erection bay, a control bay and nine main units with three service units additional in the left plant. The construction of these features by MWAK was quite similar to that of the dam as previously described, but formed areas were larger, the embedded materials (reinforcing steel, piping, conduit, etc.) were in much greater quantity and complexity. The buildings are 765 feet long, 84 feet wide and 182.5 feet high above the draft tube floor. The draft tubes for the units floored at elevation 896, slightly above the finished grade for the tailrace adjacent. The turbine discharge was divided into three segments and the draft tube forms were shaped to direct the flow evenly and with minimum turbulence. The water would exit the plant in openings some 15 by 17 feet in size. Before the flooding of the cofferdams these openings were all closed with temporary bulkheads to permit work within when construction was resumed. When finished, the powerplant foundation built by MWAK was a series of open pits—the base for the walls of the powerhouses surrounding the pits into which the draft tubes and turbines would be placed. At the top of the downstream wall and elsewhere, the reinforcement bars were left exposed to provide continuity of the reinforcement into the next lift of concrete in a future contract.

In order to accommodate the flood flow anticipated while the river was diverted through the partially constructed dam on the west bank, certain of the blocks in that power plant section of the dam remained at elevation 950 and the diverted river passed through those slots and beyond over the power plant wall. The vibration of the bars in the flowing and/or eddying water was sufficiently severe or so repetitive that many of those 1¼-inch square bars protruding 50 inches from the wall were broken off because the steel failed in fatigue. Before the walls could be extended in height, all of the bars had to be tested in tension and those not suitable had to be cut off, then extended, welding the extensions to the stubs. The vagaries of flowing water are astonishing!

Work under the contract with CBI brought the Left Power Plant building sufficiently finished to permit government forces to start installation of the turbines, and related features by late April 1940. The work was serviced using gantry-mounted revolving cranes on trackage adjacent the downstream face of the dam but supported on part of the power plant transformer deck structure. The work of CBI on the dam and other features was winding down in 1941 and the need for early construction of the Right Power House was foreseen. The contractors' personnel was being called off to the Kaiser Co. shipyards on the West Coast and the organization began to thin out. But a nucleus remained and in this climate the construction of the Right Power House was undertaken by Extra Work Orders with the contractor purchasing materials and furnishing labor and equipment to get the job done. When war broke out many of the supervisors were called to the shipyards and U.S.B.R. engineers and inspectors supervised the work directing the contractors' workmen—a most unusual arrangement. We worked hand in glove with the contractor's management and key personnel left on the job and the work was completed acceptably and

efficiently. Anyone noticing that the roof trusses on the Left and Right Powerhouses are not of the same design may be interested in knowing that the latter roof supporting structure was of steel beams salvaged from the contractor's trestle and rebuilt at the site for this permanent use.

When the power plant equipment began to arrive, the electrical and mechanical features were assigned to a separate group of engineers, inspectors, technicians, skilled supervisors and craftsmen to make the installations and do the inspection and testing for acceptance of the machinery. That subdivision reported to Mr. Darland. The generators were assembled by the manufacturer, Westinghouse Mfg. Co., but the rest of the machinery and equipment was installed by Bureau forces, the manufacturers having an engineering representative—erection engineer—present, that being a specifications requirement.

Assembly of G-3 scroll case for turbine completed. January 20, 1941.

Wicket gates to control the amount of water (and energy) to be delivered to the turbine runner. April 28, 1941.

The turbine runner G-3. These Allis Chalmers Co. turbines delivered 150,000 h.p. at full head. April 4, 1941.

Part of spider for the Westinghouse Co. generator rotor for main generating unit G-1. Mr. Fisher, pointing, indicates the six spider legs had to be shortened for shipment. September 5, 1941.

Jim Wallace (#8 Bureau) and Westinghouse crews assembled on the rotor for the G-2 Generator. October 10, 1942.

Generator G-3 complete. This unit was energized and started producing power on October 4, 1941. September 23, 1941.

The turbine pit walls of units G-8 and G-9 were removed in part to accommodate the installation of the penstock extensions through the walls for the Shasta turbines. July 26, 1942.

Carlos P. Christensen came up from Hoover Dam to take charge of the engineering aspects of the plant and James E. Wallace was employed as the master mechanic. Later, John Bates came aboard to head up the power plant operations activity. Some of the engineers in the group were: Norm Holmdahl, Raymond K. Seely, W.G. Clagett, Joe Turner, Don McGregor, Jim Green, Douglas C. Seeley, Glen Barker, Thomas, and Milton Berg. Assisting Jim Wallace were Bill Miller, Don Reeder, Doc Cook, Hansen, O'Malley, Harold St. Joer, Jack Hudson, Vic Mills, et al. Some of the operators were Ed Parmenter, Dean Cutting, Charlie Simmons and Fred Godfrey. The concrete required for the embedment of that equipment was placed by the contractor as needed. When the CBI contract was closed out in 1942, the concrete work was also performed by employees of the Bureau until 1948.

TURBINES AND GENERATORS

As soon as CBI had completed the Left Power House structure sufficiently the installation of turbines, generators, transformers, governors, circuit breakers and other control equipment installations—all items needed for the production and delivery of high voltage electric power got underway. Since my assignments were in the field of civil engineering, I had only cursory contact with these mechanical and electrical devices other than in their embedment and completion of the buildings. The work moved rapidly and by March 22, 1941, the first commercial power was delivered from the station service units. The main units—numbers 1 through 6—were manufactured in groups of three. The first main unit came on line on September 28, 1941 and the sixth on November 8, 1943. World War II

One of the Shasta Turbine scroll cases assembled. Note these turbine sections were joined by riveting. June 18, 1942.

Rotor for Shasta Generator nearing completion. Assembly of laminated ring complete. Coils yet to be mounted in the vertical slots. October 8, 1942.

was raging and the need for more power for the war industries was urgent. There did not appear sufficient time to manufacture new turbines and generating equipment to meet the need so decision was made to install two borrowed generating units (75,000 kw each) and turbines (which had been manufactured for the Shasta Dam Power Plant.) Work on that project had been suspended for the duration and these potential sources of power for the war effort were lying idle. As built, the penstocks for the Left Power Plant lay to the left of the unit centerlines while the Shasta Units were of opposite hand (and rotation). To accommodate this arrangement, large holes were excavated in the walls (turbine pits 7/8 and 8/9) and the penstock from unit 9 extended into unit 8 pit and the one from unit 8 extended into pit for unit 7. The adapting transitions were then anchored to the concrete and large concrete blocks were cast about the angle

configuration so that lateral thrust would not displace the assembly. The draft tubes were also modified to meet the constructed elements of this powerhouse. These units were brought into operation by March 7 and May 7, 1943—in time to give a boost to the war effort. When hostilities ceased, the question was soon raised as to whether these installed units should remain and new units purchased for the Shasta Plant. The wise decision was made to remove these units for return to the Shasta Plant.

The machinery was dismantled and removed insofar as possible—then removal of the concrete began. Where accessible for such procedures, the concrete was removed in large blocks by line drilling closely-spaced holes then cutting out the intervening web using a cutting tool attached to a drill rod powered by a pneumatic drill mounted on a fixed stand. The blocks so relieved, were undercut by the same methods and removed in pieces weighing up to 80 tons in size. Loaded on flat cars in the Power House and transported to the Electric City assembly yard they were dumped off into the floor of the now Banks Lake.

When the mass of the concrete had been removed, that embedding the scroll case, draft tube and other parts of the turbines were removed by workmen using pneumatically operated chipping guns, thus exposing the embedded parts without damage. When the pits were thus vacated, the walls were rebuilt and the remaining three units of the Left Power Plant installed, completing the plant with the last unit going into commercial production on April 23, 1948.

During the war and later, the installed units in the Left Power Plant were operated at up to 130,000 kw but were designed for 108,000 kw. The real story of the machines in those power plants and how they were brought into production should be told—but I was not a part of that activity so cannot do it justice.

The turbines and the generators and many of the related pieces of equipment, when installed, were substantially larger than any in service anywhere and so as might be expected, considerable debugging was encountered. The upper bearings for the generators which carried the entire weight of the rotor, shaft and turbine runner were perhaps the most troublesome elements of the machines. Bearing surfaces were wiped of the babbit contact because of insufficient oil flow over the bearings. The resultant rise in the bearing temperatures caused the machines to be stopped automatically to prevent further damage. The oiling system was improved by providing the oil directly onto the bearings by means of a pressure system and the starting routine was modified.

As I understand it, the "state of the art" when the first generators for Grand Coulee were designed was thought to have reached a degree of perfection such that fire would not occur in the generators from any cause.

Thus in the interest of economy of design, the use of carbon dioxide gas installations for automatic fire retardation within the generator housings was omitted in the first three units. However, a fire did occur, whereupon carbon dioxide gas installations were added to those units and to the rest of the units for the Left and Right Power Plants. Also to the pump motors and the generators for the Third Power Plant.

CAVITATION ON THE TURBINE RUNNERS

The turbine wheels or runners are a series of curved vanes cast symetrically about the shaft which is capped with a conical casting. The vanes are formed in complex curves and convert the energy of the water flowing down the penstock impinging against the vanes (before exiting through the draft tubes to the tailrace) into force to turn the wheel and the connected shaft and generator rotor. Cavitational forces are introduced in these wheels in operation such that the metal of the vane castings is plucked away and must be replaced. While it is possible to predict where erosion may be most pronounced, and these areas strengthened by hard facing by welding in the foundry, experience in actual use proved that repairs were needed routinely. Repairs to the runners at the dam was made with stainless steel welding rods after which the surfaces were ground smooth and returned to service.

NEEDLE VALVES — RIGHT POWER HOUSE

When severe erosion of the spillway bucket was discovered, decision was made to supplement the capacity for discharge other than over or through the spillway while dredging and repair was being undertaken. Another benefit expected from installation of the valves—as I recall vaguely—was to augment the discharge past the dam in the event any emergency caused the shutting down of several of the units in the Left Power Plant when the reservoir was below the spillway crest.

Eight large needle valves were borrowed from the Hoover Dam project and installed in the Right Power House at Grand Coulee Dam. It was to be a temporary measure and the valves to be removed as the spaces in the plant became needed for the installation of additional generating capacity. The valves were connected to the downstream ends of the riverward eight penstocks. The discharge of the valves was directed downward into the draft tube pits, providing as good streamlining of the surfaces as that configuration would permit. Since the thrust from the jets issuing from the valves was of great magnitude, here too large anchors of concrete to keep the installations from separating were provided. Also, such flow conditions engulf large amounts of air, so vents consisting of vertical pipes, ultimately 36 inches in diameter were installed to reduce the effects of cavitation. But when the first unit was tested and vibrations were observed, severe turbulence and cavitation and erosion of the concrete occurred in the throat and in the draft tube exits. There, very heavy steel castings had been installed on the leading edges of the walls between the three exit openings. These pier nose castings were streamlined to ease the flow of the water exiting the draft tubes, were some ten feet in height, and in horizontal section, were in triangular shape, the sides of the triangle approximating three feet or more on a side and at top and bottom the castings extended into the concrete about four inches. When inspection was made after the valves were operating for a sustained period, it was found that the turbulence of the water had vibrated the pier castings and then rotated them about a vertical axis, crushing the concrete embedment, so that when found, the pier castings had developed troughs in the base and ceiling of the concrete at the pier extremities, permitting them to rotate about 90° in either direction from the position as installed.

In any event, the venture while vibrant and noisy, was not a "howling success" but six valves were used when occasion demanded.

The process of removing the embedding concrete and anchor blocks was similar to that used in removal of the Shasta units from the Left Power Plants. The concrete was taken out in blocks or chunks as large as could be transported by truck.

Fortunately, demands for power soon after the close of hostilities and success in the repair and prevention of further damage to the spillway bucket permitted the removal of the last of the valve installations in 1950. In January 1949 President Truman approved the feasibility of generating units R7, R8 and R9 to complete the facility as then visualized.

RIGHT POWER PLANT

The installation of turbines and generators and related power equipment for the Right Power Plant was somewhat repetitive of the Left. Suffice it to say that authorizations and all installations had been completed by September 1951. Thus, what had been ridiculed by opponents of the project because the "jack rabbits were not potential customers," (starting from scratch) had within eighteen years become the largest producer of hydro power in the world. And furthermore, the market for it grew about as fast as it could be installed.

CENTRALIZED CONTROL OF POWER PLANTS

At the time of manufacture, remote sensing of data was not commonplace in hydro-power plants and service auxiliary equipment was used under visual, "at-the-spot" control. Nor was operation of valves and switches and greasing of bearings done other than manually. But as the means were developed or came to be available in the early 1960s, computer control of plant operating sequences was adapted to the power plant installations at the dam. Those first computers (soon replaced) used punched tape for actuating remote operations and video screens with remote television cameras and radio communication from about the plant became acceptable, dependable operating routines. When the conversion to centralized remote operation and control was completed, any generator that had been out of service with the penstock dewatered could be returned to service entirely by operation from the control panels in the central control room. That kind of operation formerly required mechanics or electricians and plant operators to manually close the penstock drain, down in the power house, then open the large valve in the 1050 gallery in the dam to refill the penstock, next from the 1295 gallery in the dam, operate the hydraulic hoist to raise the large gate at the penstock intake as preliminaries to the opening of the turbine wicket gates to start the turbine turning. Then from the 950 gallery in the power plant at the individual control panel, the unit output brought into synchronism with the system before the unit could be "put on the line." Perhaps the reader of this explanation skipped a few lines here about—and perhaps I have left out some of the steps that had to be taken in strict sequence (and each step verified) to accomplish such operations, but such was the state of the art when these machines were designed and installed. The conversion from manual to centralized remote control of the power plant and pumping plant was expensive because of the change of philosophy in management and the manner in which the plants were equipped initially. In approaching this undertaking, it was emphasized that foremost was the requirement for a *fail-safe* system and procedure. When the conversion of the plants began, the operating personnel numbered 122 and when complete, only 26 operating personnel were employed to keep the system functioning 24 hours a day, year after year. There was a lot of soul searching by me in getting into the frame of mind needed to contemplate that huge plant

Typical designs for switchyards in the 1940s and 1950s included the lattice type structures of this installation. April 15, 1960.

Initial circuits for power from the plants to the switchyard were of overhead lines supported by lattice type construction towers. Power from the west powerhouse generators to the pumps in the pumping plant was carried at generator voltage—one generator for each pair of 65,000 hp pump motors. The bus structure for those circuits extended up the incline on the face of the dam at left center. The overhead circuits were later removed for aesthetic considerations. September 27, 1967.

Routing power from the right (east) powerhouse to markets required crossing the Columbia River. At the right switchyard in the center foreground, power was routed onto overhead circuits crossing the river adjacent. Construction of the Third Power Plant at this site required abandonment of this switchyard. May 11, 1965.

and switchyards turning out such enormous amounts of energy with only five employees on duty! And I well remember the meetings I had with the power plant supervisors and operators when I explained the planned reduction in force that would accompany the centralized control envisioned!

My concerns were for the security of the plant and the risk of equipment and human imperfections hazarding the plant and the small crew. That small crew were the only persons other than a few security guards on duty for three-fourths of the time. Management, staff and maintenance personnel normally worked on day shift, 40 hours per week.

The initial "computer" installation for the plant was upgraded as newer, more effective equipment became available. I believe the conversion was vindicated in the "fail safe" results and the economies and dependable operation achieved.

My role during the period of 1953 to 1970, while Assistant Project Manager, was the alter ego of the project manager for general administration of the project, including operation and maintenance activities and as security officer at the Grand Coulee Dam Power and Pumping Plants. Responsibility for the technical aspects of the project rested with the Bureau's chief engineer from 1933 on. Also, I was the Bureau's spokesman in relationships with all hourly-paid employees under the agreement with the Columbia Basin Trades Council.

SWITCHYARDS AND TRANSFORMER CIRCUITS

Power was generated at 13,800 volts and transformed to 115,000 and 230,000 volts in transformers located adjacent the rear wall of the power plants. The transformer circuits carrying the energy to the switchyards and the switchyards were built using latticed type structures

Before construction of the Third Power Plant could be started the entire right switchyard service had to be relocated to the Left Switchyard site. The entire yard was redesigned using more aesthetic treatment and with capacity increased to service the output of the original two powerplants. Third Powerplant construction in progress in distance. August 2, 1973.

Circuits from the power plants to the switchyards initially included towers supported on the power house roofs. August 13, 1959.

which was the standard mode of the day. That type of structure perhaps used less steel and labor was available and "it had always been done that way." The 115,000-volt switchyard which was located on the left abutment at about elevation 1625 served local and station needs and two transmission lines to Spokane. A 230,000-volt switchyard was located at elevation 1700 also on the left abutment where space was available and terrain permissible without excessive excavation and filling. In 1947, that switchyard served six outgoing lines—two to Spokane and four to the coastal cities. The switchyards were built in stages as new designs for circuit breakers and switches came along though the drilling and blasting of the rock was kept ahead of the equipment installations when possible. The Right Switchyard was constructed near the present location of the Third Power Plant at about elevation 1174, under Specifications No. 2075 by the M/K-P/K Cos.

The switching and protective devices for control of the energy systems of the size and complexity of that contemplated for the dam were in a state of development because of the very large interrupting capacity required. Thus, construction work and installations went hand in hand adjacent live high voltage circuits requiring extra precautions for safety.

In the mid-1960s when the problems associated with the Third Power Plant were being reviewed at the dam, the plan for converting the transformer circuits to oil-filled cables through the dam was under discussion and I well remember the retort I got when I suggested that the floor of the dry dock would make a strong foundation for a tower. "What would that look like?" Aesthetics were vital.

The Third Power Plant required the abandonment of the entire right switchyard and the relocation of the transformer circuits. Also, the trnsformer circuits from the Left Power Plant were also replaced with oil-filled cables. The cables were routed through tunnels through the rock and in the galleries of the dam. Thus all of the old lattice type construction and angle iron tower structures adjacent the dam were removed. The 230,000-volt switchyard on the left bank was completely rebuilt using streamlined, low-profile designs with the yard enlarged sufficiently to carry the entire capacity of both Left and Right Power Plants. That rebuilding, too, was accomplished with minimum disruption in the flow of energy.

Overhead circuits were again on the scene in the early eighties after a disastrous fire destroyed some of the oil-filled cables from the Third Power Plant.

PUMPING PLANT DISCHARGE PIPES

When the pumping plant discharge pipes went into service, immediate attention was focused on the pulsing vibration of the pipes apparently in harmony with pulsation in the discharge from the 1600 cfs pumps. The movement was sufficient to be visible from some distance and considered unsafe to continue indefinitely. To remedy the condition, ribs or flanges were welded to the exterior of the pipes in the unembedded portions below the outlets. Another difficulty encountered at startup was the failure of the siphon breakers to respond to the cessation of pumping with the result that the water in the canal was drawn back into the discharge pipes and back through the pumps into Lake Roosevelt. This required modification of the siphon breakers for quick and adequate response with sufficient air intake to interrupt the reverse flow.

WAREHOUSES

Two large warehouses were constructed in the shop area between the dam and Grand Coulee. The design included filling on the existing site to bring the floors up to truckbed level. The site had been levelled using waste material from the road and railroad and the dam excavations in the early contracts (1934-1936). The areas so leveled had been used for storage of materials prior to construction of the warehouses and seemed to have stabilized. I am uncertain whether bearing tests were made prior to the design and construction of the warehouses, but in any event, substantial unequal settlement of the buildings occurred shortly after completion in 1949. Also, settlement of the unconsolidated fill was a recurrent problem when the highway through the shop area was relocated to skirt that area and it too required periodic restoration of surface. But now, after nearly 50 years, the fills seem to be stable even though irrigation of the trees and shrubs along the highway occurs. But it took a lot of years for the fills to stabilize on their own accord.

CUSTODIAL FACILITIES

Mention has been made heretofore of the drainage holes in the foundation of the dam. Also, vertical porous drains were installed at ten-foot centers located about fifteen feet upstream from the longitudinal gallery system. These drains provide a means of escape for any seepage water that might enter the concrete of the dam from the reservoir. These drains also discharge into the gutters of the galleries. Controlled pipe drains also discharge into the gutters from the outlet gates' lower bonnets and operating sealing control systems. The water from all sources coming into the galleries is collected into three sumps located in blocks 31, 53 and 64 which are at the ends and midway of the spillway. Pumps lift the water up above tailwater level for discharge through the spillway retaining walls. Water from the control of the drum gates also discharges through large pipes opening through the training walls.

Access throughout the dam and into the pumping and power plants is gained through the elevators and the stairways. Adequate lighting for access is permanently installed and is augmented where needed for maintenance work. Until the Third Power Plant started producing power, the temperature in the galleries in the dam was near 45° F which was the temperature of the concrete after cooling operations were finished. Since the oil-filled cables were installed, ventilation systems were added so that the air in the galleries is warmer than previously.

CHAPTER X
WARTIME ACTIVITIES AT THE DAM

SUPPORT FACILITIES

The Grand Coulee Dam and Power Plant were essential industries for the *War Effort*, supplying electric energy to the aluminum industries and to the shipyards of the West, so number one priorities were allotted to accomplish the needed work of operations, maintenance and installation of "more power." But lumber was in scarce supply and the contractor's mixing plants had been removed when the work under that contract was finished. The Bureau forces then constructed a sawmill on the reservoir shore where saw logs towed down from the timberlands to the northeast were sawed as needed. To expedite the preliminary preparation of turbine and other parts, an assembly building was constructed where welding, fabrication, sandblasting and painting could be accomplished in any weather. Maintenance shops for servicing and repair of trucks were also built. A mixing plant adequate to supply the needs for concrete to embed the turbines and for the generator foundations was built near the stockpiles of aggregate produced for that purpose before CBI dismantled and removed its plant—and sand was in plenty!

On January 10, 1942, a 33-man contingent of the U.S. Army arrived, took up quarters in Mason City (barracks in the camp) and performed guard duty with sentrys at strategic locations of the dam, power plants and switchyards, supplementing the Federal Guard force that controlled access within the facility. The soldiers were present because of the possibility of an enemy thrust to the area after the Pearl Harbor disaster the previous month. But they were not well armed for that contingency, should it occur. No efforts of sabotage of the dam and power plant and switchyards came to light—I recall none such at any time during the conflict.

Then, had there been any, and they were "caught in the act," they may have been shot. Now, in 1986, I suppose any saboteur caught in the act would plead mercy, would be taken into custody, provided a good, well-paid attorney at government expense, would be well-housed, finally tried, and—if found guilty—would appeal. Lastly, if appeal failed, he might even have to serve a few additional days, though in comfort, well-housed and fed!

Enough workmen, engineers and inspectors stayed on the job, and many women—wives of workmen and men in the service—were hired and helped to keep the installations and operations moving with little delay.

WARTIME INSTALLATIONS

First on the list of "to do" items when the war engulfed the United States was the duty to keep the dam and the power plant safeguarded and in full production. So certain protective works were added to deter any would-be saboteurs and guarding of the premises became a bit more sensitive. The work on the Right Power House building was being constructed by CBI under an extra work order and that work continued, the building being finished by January 1943.

In the Left Power House the turbines and generators and related equipment continued to be installed, tested and put into service as rapidly as was possible with the means at hand. Main generator G-3 went on the line on October 4, 1941 and was followed by five additional main units on January 29, 1942, April 7, 1942, August 9, 1943, November 8, 1943 and February 12, 1944.

Two turbine and generator units from Shasta Dam were delivered and installed and put into service by February 25 and May 7, 1943.

Of course the installation of turbines manufactured by Allis Chalmers and generators manufactured by General Electric was not the whole story. There were the bus connections to the transformers, the transformers, the lightning arresters, the circuits to the switchyards complete with switches and circuit breakers with connection to the outgoing lines.

All of those activities, though predominately in the hands of the electrical and mechanical sub-division, all required support and concrete construction for embedment, bases and foundations. So coordination was especially important to prevent any delay.

The condition of the spillway bucket was yet to be seen but when the cofferdams and cribs were being removed in the final cleanup of the MWAK features appreciable amounts of debris were removed from the spillway bucket. This was a nagging concern, so an inspection of the spillway bucket by divers was undertaken in 1943. Consideration was then given to the need and timing of repair of the damaged bucket. And in the chief engineer's office, Bob Sailer and others were designing a floating, submersible caisson with which to repair the damage "in the dry." As soon as the dimensions of that device were assured, we got a "go ahead" to build a drydock to house and service the monster. That was in the early spring of 1944.

Meanwhile, the hydraulic model (scale 1:60) was constructed and tests run to determine the limits of excavation of the river mantle downstream from the bucket. James W. Ball, a hydraulic engineer from the chief's office, came on May 18, 1943 to the project to direct the testing activity.

Concurrently, the installation of the needle valves in the Right Powerhouse and the testing and modification of these installations was in progress. When I talked to Bert Hall just before his 93rd birthday on July 2, 1986, he reminded me that the job at the dam during the War was the most challenging and difficult of all!

Dredging of the river channel was started in 1943 and continued during the low stage of the river each year until 1950. Throughout the period of the War and shortly thereafter, delivery of materials and equipment lagged and improvising was the order of the day. I recall that most—if not all—of the control panels for the turbines and generators were initially made of plywood in order to not delay the start of the units into production.

WORLD WAR II ENDS

When World War II was nearing the end, attention was focused on the hiatus in the economy while the country was changing from a wartime all-out effort to a normal pace of life and industry. Decision was made to *Gear up immediately!* to get work started on the completion of the dam and facilities and soon to start the construction of the irrigation works. Since the work at the dam had been pushed along during the War to produce power for industry, the Bureau had a nucleus of a skilled work force which could be readily expanded to provide employment for many who were suddenly out of work as the war industries were shut down. So the employment rolls were increased and decisions were made to complete the work at the dam with Bureau forces instead of by contract. Plans began to take shape and the

construction pace picked up to meet that objective. And the schedules for accomplishment of the work were tight! I recall a meeting we had with Mr. Banks and others to consider the needs of the job to get the pumping plant into service and deliver water to the lands to be irrigated on the Columbia Basin Project. When asked what was needed to meet the schedule, Bert Hall—perhaps with tongue in cheek—replied, "We need the complete, detailed design drawings, the materials and supplies, and the manpower by [he mentioned a certain date.]"

Mr. Banks retorted, "Well, Bert, you know you won't get that, but you will have to do it anyhow!"

From the Hanford Atomic Works, we brought in equipment of all kinds and supplies as well, and from the shipyards, I selected the large whirley cranes that would be needed for construction of the pumping plant and the program got underway. A large addition was added to Mason City to house workers and some trailer camps were now set up for the same purpose. And excavation of the feeder canal from the outlets of the pump discharge lines to the north dam of the equalizing reservoir got underway.

But soon the contracting fraternity became aware of the government practice of doing construction work by "force account" as it was called, and no doubt made their objections known. In July 1947 the Congress of the United States took action to cut back construction work performed by government employees—henceforth all construction work would be performed by contract. And for the long range, I think it was a good decision. With a contract for a certain work, the work must be described, the quantities and quality determined and specified, the method of payment and the amounts fixed by bid or negotiation, and the time for completion and cleanup of the worksite fixed. The latter was particularly a problem with "force account" activity—there were too many unfinished jobs and not much leverage to get them cleaned up! But short range, the decision to stop "force account" was a bit heartless—and resented. In the last half of 1947 the reduction in force caused the layoff of 1300 wage rate employees—707 or 78% of the civil and structural division and 464 or 60% of the electrical and mechanical division, and others. And those decisions—who must go—were not easy or pleasant! Those people had been encouraged and some sought out to come to the dam to help us! Now we were sending them away!

CHAPTER XI
SPILLWAY BUCKET

UNDERWATER INSPECTION

It has been said that all overflow dams have sore toes and Grand Coulee Dam was no exception. Of the possible choices for energy dissipators from the spillway discharge the surmerged bucket design was selected and it has been very effective. But the failure to anticipate the destructive forces created in the disposition and excursion of riverbed materials carried into the bucket resulted in severe damage. The material in place and waste from excavation was permitted to be leveled off at elevation 895, five feet below the peak of the downstream limit of the bucket. During construction bed load of the river upstream of the diversion channels through the dam may have been picked up and carried into the spillway bucket. Also after the cofferdams were removed and while the flood waters were being diverted over partially constructed blocks of the dam and through the outlet works, great eddy currents were present in the pattern of the rushing river. The magnitude of such and the apparent velocity coupled with the depth of the water combined to subject the concrete of the spillway bucket to large deposits of material from the riverbed and to abrasion while moving over the surfaces. Observation of those flow patterns raised a feeling of concern. From the lowest gallery in the Right Power House, where no machinery was operating, I could distinctly hear the pulsing or intermittent rush of heavy gravel and cobble.

In the spring and fall of 1943, inspection of the spillway bucket was made by divers. Hard hat or helmeted with pressure suits was still the normal means for underwater inspections at depths of 70 feet—scuba diving had not yet ascended. These inspections which covered the entire spillway bucket and the riverbed adjacent disclosed that the bed materials from the river bottom had been deposited in the spillway bucket in substantial quantities and consisted not only of sand, gravel and cobble, but pieces of sheet piling and all sorts of metal parts from construction activity. And this material had severely eroded the lip of the bucket at a location where the migration of material to or

from the bucket had been concentrated. Also, erosion along the horizontal contraction joints in the bucket concrete had extended well below the erosion nearby.

The assumption that such an erosional pattern occurred because round rocks were traveling forth and back along the formed line was confirmed when one of the divers noticed such actually happening while he was nearby. It was most disheartening to us to find the bucket in such a deplorable state—we had had it "as clean as the kitchen floor" before removal of the cofferdams. In the inspection of the bucket, the divers were employed on an hourly rate and reported what they saw to the inspector, myself and others, on the diving barge by means of a telephone.

From the descriptions reported, the conditions encountered were recorded and from those notes, drawings were prepared of the entire

After removal of the downstream trestle, the voids or blockouts about the leg locations were filled with concrete. All debris was removed from the entire bucket section in readiness for flooding of the cofferdammed area. Openings in the face of the dam are the lower set of 102-inch outlets. Spillway training wall at far end of bucket was still being constructed, but other concrete work in the cofferdammed section of the spillway was now at the limit for MWAK work. November 24, 1937.

surface condition. This inspection was serviced with a barge-mounted crane and controlled and moved along the face of the dam by hoist cables anchored to the dam and to shore. Later when these surfaces could be visualized after dewatering of the floating caisson it was found that the divers had reported the depth of erosion to within a fraction of an inch using only their experienced eye and the judgment of their construction experience with very few actual scaled measurements.

Those divers, brothers Colin and "Spud" O'Donnell made the survey an event to remember. Lastly, I should comment on the effectiveness of the deep pool as an energy dissipator. On one occasion the divers were moved onto a section of the bucket where the spillway gates had just been raised but had been discharging some 5,000 cfs each for several days. The divers reported that algae was present on the undisturbed gravels lying on the surface of the deposit in the bucket.

The chief engineer forwarded my report, dated April 15, 1943 to consulting board member Dr. W.F. Durand for recommendations. He replied on May 7, 1943. He considered some of the erosion of the concrete in the spillway bucket to be caused by cavitation—as had we in the field. He recommended model tests, studies for repair of the damage, provision of by-pass facilities in the Right Power House, and removal of river bottom materials downstream from the bucket in the future.

FLOATING CAISSON

The chief engineer's office as well as we project personnel were alarmed at the conditions found and determined that corrective measures must be undertaken as soon as possible. Designs were prepared and models were made to test the various features of the equipment needed to undertake the repair. The resulting plan and equipment then began to take form. A steel caisson would be constructed that would be supported on an integral barge. It would be moved into position by large, shore-mounted hoists. Then it would be anchored to the face of the dam at the center line of the block to be repaired using an A-frame attachment pinned at the bottom of the legs to the caisson itself. The caisson was then to be lowered to the spillway bucket and set onto preconstructed seats. The caisson was to be equipped with large, heavy, rubber seals attached along the entire bearing surface. The caisson had overall dimensions of 60 feet wide,

An operating hydraulic model (scale 1:60) of Grand Coulee Dam and the river channel downstream was constructed upstream from the Pumping Plant. Water was discharged through the outlets and over the spillway of the model to simulate operations and to determine the effects of the induced currents upon the riverbed materials. The plastic model of the floating caisson was maneuvered in the currents to determine tensions required in the control cables for the puller machines (large hoists) used to position the caisson while enroute to or from the bucket of the dam. August 27, 1943.

119 feet long, and by 100 feet high. It permitted work performance with river level at elevation 945. It spanned an entire block and provided a working space and headroom and permitted exposure of the downstream face of the dam up to elevation 905. Within its four large shafts were an elevator and ladders, pumps for dewatering and open shafts for the passage of materials out of and into the work area. And it "worked like a charm!" At the first setting, I recall going down into the bottom of the spillway bucket three hours after dewatering started, with only a slicker and oxford height rubbers for protective clothing and didn't even get my feet wet. But that is getting ahead of the chronology!

In order to construct, operate and maintain the floating caisson, a drydock was to be constructed on the right riverbank where the contractors' shops had been located. Also, means had to be provided for construction of the seats upon which the caisson would rest in the eroded portions of the bucket. and the Hoover Dam needle valves to be installed in the Right Power House would be used for emergency discharge when necessary because of curtailed or interruption of power production in the Left Power Plant, or when flows over the spillway unduly interferred with the bucket repair.

But the 1947 and 1948 flood flows overtopping the drydock caused considerable damage to the floating caisson which required repair before use. Also the planned use of Bureau forces to make the repairs had to be scrapped and a contract was awarded to Pacific Bridge Co. for that service.

DRY DOCK

The design for the drydock to accommodate the floating caisson provided for a reinforced concrete structure set on bedrock with a floor elevation of 905. Upon the floor would sit pedestals upon which the caisson would be assembled and sit after each season. The drydock must extend sufficiently above low tailwater levels to permit a sufficient working season and to permit recurrent use of the caisson, it would have to be stored during highwater. The location selected for the drydock was quite exposed to the seasonal floods, but it was favorably located with regard to the hoist locations needed to maneuver it to and from the spillway. The top of the drydock then was set at elevation 959.5 above which the area within would be submerged, having earlier been filled with water. The river would be excluded from the work

Percy Pharr and "Big Bill" Walker operating the model of the floating caisson to determine optimum criteria for control of the sinking process while Bert Whiting looks on. October 29, 1945.

The caisson for repair of the spillway bucket under construction in its circular graving dock. January 27, 1947.

The floating caisson emerges. Floating caisson leaving the drydock enroute to the spillway bucket. August 24, 1949.

Site preparation for construction of the drydock for the floating caisson. The cutting edge of steel to be installed in 16 sections weighed 45 tons and was fabricated in project shops. Assembled at the toe of the sloping surface shown here the cutting edge was 163 feet in diameter. May 6, 1944.

The drydock was constructed as a caisson and after suitable height had been attained, the mud-sills (timbers) supporting the cutting edge were removed and the structure lowered into the overburden to bedrock by excavating the supporting earth. July 21, 1944.

The drydock support was gradually removed by excavating the clay within using dozers to push the material out into the open where it was removed by clamshell buckets and hoisted topside for disposal. October 11, 1944.

147

Drydock structure complete, awaiting installation of mitre gate. Approach channel yet to be excavated. March 28, 1945.

Opening the drydock gate. Opening the circular drydock to permit the floating caisson to move out into the river. This shows the front arch segment being tipped into the channel excavated to receive it. This segment weighed 1700 tons. March 19, 1949.

space within the drydock by means of mitre gates with a height of 54.66 feet and a clear opening of 65 feet. The gates would be submerged at all times when the river flow exceeded 200,000 cfs.

The drydock was to be constructed as an open caisson and sunk to bedrock by excavation within. When the caisson for the drydock was in place and completed, ready to be flooded, the wall of the structure in front of the mitre gates would be tipped out into the bottom of the channel which served the going and coming of the floating caisson.

The site for the drydock was cleared and leveled down to elevation 965 at which point the excavation continued only on the work area outside the 163-foot-diameter of the dock and a truncated cone from that diameter was left intact to serve as the form upon which the sloping surface of the structure above the cutting edge would be cast. The cutting edge was fabricated locally from salvaged, heavy steel I

beams, the sections having been shaped to the curve of the periphery of the dock before delivery to the site. The cutting edge then was assembled on mud sills of 4 by 12 timbers about five feet in length, spaced at about 20-inch centers around the periphery. The structure was to be launched or "cut loose" when it had a height of 33 feet and a total weight of nearly 14,000 tons. The reinforcing steel in the ring section of the dock was plentiful it seemed, and one of the old timers on the inspection crew beseeched me to request a design change to omit a portion as not necessary, but I did not concurr—it was a very large structure and it had a long way to go to bedrock. Of that—more later!

The interior diameter of the caisson section below the floor level was 131 feet, the walls being 16-feet thick. The reinforcing for the mitre gate anchorage was much more dense in spacing.

When the concrete in the "launch section" had set and cured sufficiently, excavation within the confines began using a dozer to move the earth from the walls and from the conical area section of earth under the sloping surface. Also the mud sills were cut off from the outside flush with the steel cutting edge—at the outset every fourth one was cut and as excavation inside continued then every other mud sill was cut off and then another round was cut off the crews working around the periphery at points diametrically opposite to avoid unequal support when movement would begin. The suspense was beginning to mount as we all were amazed to see that large mass of concrete standing there with what appeared to be greatly insufficient support. I was not present when the caisson made its initial move—but it was accompanied by sounds such as that of a volley of rifle fire and the structure dropped about five feet into its supporting earth. I was in the shack at the top of the slope when the "old timer" dashed up the stairs red-faced and panting, and greeted me with, "My God! I'm surely glad you didn't let us remove any of that steel!"

Cracks about 3/16″ wide at about ten-foot intervals had opened throughout the circumference of the structure when it dropped. Excavation continued inside the ring of concrete with additions in height as the structure continued to bury itself. When settlement got a bit uneven, excavation inside the pit was slowed on the low side until it again was on an "even keel." When the cutting edge reached bedrock at a high point in the very uneven rock surface, a bit of crushing of concrete occurred but it was localized. The rock point was removed by blasting to achieve the desired depth. Then piers were constructed under the caisson section just beyond the sloping surface above the cutting edge. Since the bearing on the rock was quite localized and to avoid the risk of a "blow in" these piers were sunk to bedrock as small caissons with short sections of concrete rings for support. The rings were then filled with concrete to support the structure against any further settlement. Concrete was placed under the caisson walls and four 10-foot square concrete columns were constructed to support the 16-foot-thick concrete floor and the pit then was filled with pit run gravelly material up to elevation 889. The floor was then constructed and the walls completed. Above the floor level, the walls tapered to a thickness of five feet at the top. The portion of the wall on the back or shore side was raised to support a service road to the structure at elevation 970.

The mitre gate was then installed and heavy steel cables were attached to the inner surface anchoring it to the concrete of the drydock.

However, when the subsequent flood had receded and the gate again was visible, it surprised all to find the gate open and resting against the concrete outer wall. One of the large pins holding the gate at its top had risen out of the bearing.

It was apparent that the steel cables with turnbuckles which were installed to keep the mitre gate secure during submergence in the flood failed either by over-stressing or, more likely, by fatigue from the repetitive stressing. The pin that rose out of the bearing had caused its keeper to fail when the gate, now free to swing, swung excessively.

Before the next submergence of the drydock, very heavy steel beams were installed rigidly anchoring the mitre gate to the drydock floor.

The floating caisson was assembled and tested for operational sufficiency and the portion of the drydock cylindrical section opposite the mitre gate opening was tilted out and dropped into the bottom of the approach channel which had been excavated to receive it. Before tilting the circumferential reinforcement steel on the outer and inner bands and from the vertical steel at the floor of the drydock entrance was cut off so that the concrete section could be removed within the construction joints built into the structure initially.

In preparation for submergence during storage in the highwater period, all hoists and machinery were removed and all hatch covers were bolted down. It too suffered from the pulsing of the over-topping waves and surges from the spillway and when next viewed, the hatch bolts had been broken and the hatch covers displaced. These failures too appeared to be fatigue failure.

At the beginning of the effort to construct the drydock, a 40-ton cable way was purchased and installed with the head tower on the west bank of the river and the traveling tail tower on the right or east bank. The cable way was also used to service equipment being used in the dredging of the river and other operations in the river below the dam. With the cable way, loads could be picked up or deposited on either side of the river. Stiff leg derricks and crawler-mounted cranes were used for excavation, placing of steel, forms and concrete for the drydock. The derricks were removed before the dock area was flooded and the cable way serviced the dock thereafter.

The rubber seals had sufficient depth and resilience to seal the floating caisson to contact to the face of the dam and the spillway bucket but where the erosion was more than about 1½ inches in depth, it was deemed necessary to construct a support of some type upon which the caisson could be set and sealed to insure that the work within would not be delayed. Decision was then made to construct a steel frame shaped to the design profile of the spillway bucket and to construct the seat of prepakt concrete. The seat form provided the upper surface of the seat and the sides were constructed of timber attached to the frame. Small depths between the eroded concrete and the form were closed by caulking. The underwater work, of course, was performed by divers.

The seat frame was designed (at the field office by Ernie Hill) of the same lateral dimensions as the floating caisson. Structural stability for the bottom plates against which the seat would be cast was obtained using welded truss members and the uplift forces created by the intrusion of the grout were countered by use of external loads (pig iron in cribs) distributed along the periphery of the frame. The frame, and later the floating caisson, was attached to the anchors on the face of the dam. Lateral position was controlled by reference to the contraction joints of the block under repair. Normally, it was found that the sloping surfce of the bucket support block downstream from the lip of the bucket supported the frame at about the desired elevation. However, towers were provided and used to verify the desired elevations of the frame. The towers extended above the water surface and elevations transferred by surveying methods assured accurate determination of the elevation of the seat form at the bucket. Divers using jack bolts attached to the frame made final adjustments before the ballast crates were lowered into place. The space within the form was then filled with selected gravel through access hatches in the skin plate of the form. When filled, grout was then forced into the form beginning at the lowest elevation in the bucket and progressing up the slopes from the bucket centerline until the return from the vents indicated complete filling of the form. Low pressure grouting was essential to avoid displacement of the form which was five feet in width along the entire periphery of the seat. The seats of pre-pakt concrete so constructed, had good bond to the surfaces of the eroded concrete. Such

having been cleaned of objectionable material by the divers, prior to setting the form in place. When the floating caisson was lowered onto the seats and dewatered, there was no displacement or loss of seal due to the pressure of the water from without. In fact, some of the seats remained in the bucket through several flood seasons without damage. Because these underwater methods were so successful, it was decided later that future repairs could be performed entirely by such means and the floating caisson and drydock disposed of.

Now, in 1986, the seats are still present in six of the blocks which have not been repaired. The divers report that shallow portions have been dislodged, but for the most part the seats are relatively undamaged.

REPAIR OF CONCRETE IN SPILLWAY BUCKET

When the floating caisson was in place, the surface of the concrete was examined to determine the extent of the erosion. In the 11 blocks that were repaired the measurements showed that about 700 cubic yards of concrete had been lost by the erosion from foreign materials carried into the bucket by the turbulent waters. The maximum depth of erosion noted was about five feet, measured normal to the surface. Where the erosion exceeded one inch in depth, the concrete was removed to a depth of 18 inches before installing steel bars anchored into the concrete and then restoring the original surfaces with concrete. Surfaces having less than one inch of erosion were carefully ground to a smooth surface. Concrete was delivered into the bucket for repair through the service shafts of the floating caisson. Steel screeds were used to shape the bucket to its original curvature and the steeper slopes were formed to confine the concrete. The floating caisson was used over a period of four years, 1949 to 1952, in repair of the 11 blocks requiring 1800 cubic yards of concrete. The surfaces of the replacement concrete were covered with a white curing compound. Divers reported that the curing compound was still visible on the surface and only slightly worn after two or more flood seasons but it is no longer noticeable the divers report. But trowel marks are still visible on the surface of the repaired blocks.

The success of the underwater construction of the seat forms using prepakt concrete and the removal of the source of the erosive materials from the riverbed permitted deferral of further repair to the spillway bucket until the need becomes evident.

It was not possible to use the floating caisson for repair of the damaged areas in Blocks 31 and 64 at the extreme ends of the spillway so the bucket surfaces in those blocks were shielded with steel plates covered with a rubber coating vulcanized to the steel. This option was used because the wear being experienced in the bucket was from abrasives grinding the surface and the rubber has much greater resistance to abrasion. The wear plates were installed by divers, the plates being anchored to the concrete by bolts secured into holes drilled in the bucket concrete. The space under the plates was then filled with grout. Unfortunately, these wear plates have not proved to be as successful as we had hoped or expected and the divers found that some of the sections of steel plate are missing and the rubber coating has been torn off other plates—but some are still in good condition. It is possible that the damage to the wear plates may have been caused by the milling about (in the water cascading down the face of the dam) of large water-soaked logs such as are occasionally found in the bucket at the time of inspection. Such "sinkers" are occasionally lost from log booms at the saw mills upstream from the dam and migrate downstream, finally becoming entirely saturated and submerged. The adequacy of the bucket seems assured, providing of course that diving inspections are made and dredging performed to remove potential sources of abrasive materials from possible entry. Since the repairs were made, annual inspections have been made by the divers.

Scuba divers are now doing the inspections and their most recent (1985) report disclosed little change in the condition of the bucket since 1952. Some movement of material (loose rock and gravel) on the bedrock downstream from the bucket has occurred, but I was informed that such materials have not entered the bucket.

In 1943, when the first inspection of the bucket was made by divers, scuba diving had not arrived. Neither were satisfactory underwater cameras, insofar as we could determine. One "homemade" camera was used, but the picture was not as good as the description from the eye of the diver. Years later cameras and lights suitable for underwater photography were developed, permitting accurate continuous video tape recordings of the scenes.

With the construction of the Third Power Plant and with the success of the measures taken to remove the source of wear in the bucket of the spillway, all of the special equipment for repair was dismantled and disposed of—the cable way, the drydock, the floating caisson, the large maneuvering hoists from the riverbanks, the seat frame, the large derrick barge and the face caisson are now only a memory of a most unusual undertaking.

The contractor, Pacific Bridge Co. (specifications no. 2475) undertook this work with dispatch and Messrs. Jim Ginnella, the contractor's manager, Charley Bisordi the superintendent and George Noe the engineer on the job all worked with us in full cooperation in this successful venture.

Floating caisson, attached to spillway face by A-frame. March 2, 1949.

The inspection of the spillway bucket and the adjacent river bottom by divers disclosed that the eddy currents in the river below the spillway carried the riverbed mantle into the bucket, first having created large deposits of such in ridges extending for some distance downstream. It was obvious from these observatons that the principal cause of damage to the bucket was the grinding of the foreign materials carried by flowing water. Decision was made to remove all earth and loose rock or other material from the bedrock of the river channel for a distance of 200 feet downstream. Later, after model tests and observations by divers, the limit was extended to 300 feet. The material was excavated by Bureau forces and later by the contractor, Pacific Bridge Co. Barge-mounted, steam-powered cranes using clamshell buckets excavated the material which was placed on barges and deposited in the river downstream from the highway bridge. Each barge was unloaded by use of a small dozer with which the material was pushed over the side. This dredging operation was performed during low water over a period of years, 1943 to 1951. The bedrock was exposed and only a shallow covering remained.

Scuba (self-contained underwater breathing apparatus) diving teams were later organized in the Bureau of Reclamation for underwater assignments. The inspection of the spillway and other features of the dam has been performed by scuba diving teams for several years.

Presumably, inspections by divers will disclose when and if further dredging will be necessary. I am not familiar with the effect, if any, that the construction of the Third Power Plant and its operation may have on the movement of materials into the spillway bucket, but certainly the currents in the waters below the dam and adjacent the spillway have been greatly changed.

Dredging river bottom deposits to prevent entry into the spillway bucket. This work was first undertaken by Bureau forces—but later performed by Pacific Bridge Co. under specifications 2475. The floating caisson was erected within the drydock upper right. Barges disposed of excavated material downstream from the highway bridge. February 23, 1949.

CHAPTER XII
IRRIGATION WORKS

PRIMARY FACILITIES

While I was busy as a "dam builder" I was barely aware of the activities of Nat Torbet—but such would have a big influence on my career. Nat was living at Coulee Dam and planning for the development of the irrigation elements of the Columbia Basin Project. Until repayment contracts were executed with the owners of the project lands, the construction of the features to bring water there could not be undertaken. I do not recall the signing of the repayment contracts as an event of which I took much notice, if any. But with those contracts with the three Columbia Basin Irrigation Districts on October 9, 1945, the pumping plant, the feeder canal and the two dams at the limits of the upper Grand Coulee (to form the Equalizing Reservoir—Banks Lake) got the "go ahead signal!" Those features were assigned to the Coulee Dam Division for construction—a part of Bert Hall's reponsibility. The remainder of the irrigation features were assigned to the supervising engineer at Ephrata, Mr. H.A. "Hap" Parker, who came to the project in 1939.

As mentioned herein, construction of the pumping plant and the feeder canal was started as a "force account" activity but that was called off and the completion was by contract—as was the construction of the North and South Dams (later to be Dry Falls Dam). Those dams were constructed by J.A. Terteling and Sons for the North Dam and Bair and Crick for the latter. H.P. "Pat" O'Donnell was the resident engineer on those dams.

With the prospect of the coulee becoming a reservoir attention moved to the need to abandon the railroad and relocate the highway in the coulee. Negotiations with the State Highway Department moved along well—we dealt principally with Mr. James Davis, the Highway Department construction engineer. Agreement was reached on the typical design for the highway. It would be ballasted to a depth of 27

inches as I recall and for a width of 34 feet but surfaced for a normal width. The highway was intended to carry the heavy loads anticipated in completing the powerplants using the wide, solid-tired, 200-ton capacity "Boulder" trailer. Maximum grades would be 6% but the terrain was such that Pat O'Donnell our locating engineer was able to locate it with 5% maximum grades. The so-called "million dollar mile" where the highway rises up over the coulee wall was required because the coulee bed at that location was of saturated silts that would not support the high fill required.

The railroad was temporarily relocated around the North Dam so that it could be operated until the time for filling the reservoir was imminent. Then the railroad and old highway were both abandoned. The rails and fittings of course were salvaged. Finally at the beginning of the Third Power Plant construction, the railroad bed down the hill from the dam was utilized when an additional lane was provided for slow, uphill traffic on the highway from Coulee Dam.

Tork Torkelson (Bureau) assists as Jacob Weber signs the Repayment Contract on behalf of the Quincy Columbia Basin Irrigation District. August 31, 1945.

Tork Torkelson assists Loen Bailie as he signs the Repayment Contract on behalf of the South Columbia Basin Irrigation District, H. de Schepper, district secretary. August 31, 1945.

Mr. Torkelson also assists Don Damon signing Repayment Contract on behalf of the East Columbia Basin Project, C.H. Brittenham, district secretary. August 31, 1945.

Secretary Ickes signing the Columbia Basin Project Repayment Contracts at his office in Washington, D.C. Standing: Representative Walt Horan, Governor Mon C. Walgren, Senator Hugh B. Mitchell, Representative

Hal Holmes, Senator Warren G. Magnuson, Frank T. Bell, member Col. Basin Commission, Harry W. Bashore, Commissioner of Bureau of Reclamation. 1945.

CHAPTER XIII
THE THIRD POWER PLANT

GROWTH OF POWER DEMAND

In the early 1930s planning for the dam, the initial expectation of capacity that could be justified for the site was 15 units of 108,000 kw. However an additional three units were found feasible by the time the specifications for the low dam was issued. And the demand for power grew faster by far than anyone had predicted or planned. So, in the early 1950s another study was undertaken to determine the feasibility of constructing additional hydroelectric installations at the dam. The conclusion in a report dated 1954 was that four additional units each of 108,000 kw could be economically justified on the basis of the then-economic factors and projections. However, the growth in demand continued and the desirability of hydro-power for peaking purposes increased relative to thermal units (either atomic-, coal- or oil-fueled). Also, larger units were coming into service and the interconnected transmission systems were able to handle the "shock" of sudden, larger load losses from natural or switching responses. And centralized and remote operations of plants was coming of age.

CANADIAN TREATY STORAGE

About that time a group of utilities in the Northwest proposed that they undertake to finance the construction of Canadian storage. An agreement would be negotiated with Canada for the construction of additional storage dams in Canada on the headwaters of the Columbia for release of the summer runoff during the fall and winter months to firm up additional power availability in the United States. With this prospect, the feasibility for additional capacity at all of the dams on the Columbia River was cast in a new and brighter light. Near the end of World War II, the United States Government had opened formal negotiations with the Canadian Government to determine the feasibility of developing upstream storage in the Canadian drainage of the Columbia River.

Near the conclusion of those negotiations a delegation of Canadian officials, 17 in number, headed by Mr. Paul Martin, Secretary of State for External Affairs and Mr. Arthur Laing, Minister for Northern Affairs came to the dam.

I regret that I was not a very good host for the group since we had very little advance notice of the visit or any information as to the full purpose of the visit or their specific interests. Mr. Martin and his aide rode to the dam with me, as did Regional Director Nelson. Enroute Mr. Martin told me that he wished to go swimming in the Columbia—Lake Franklin D. Roosevelt. That was a new problem as I saw it, and I tried to dissuade him. Finally he stated that I was not very encouraging about his swim. When I replied, "I don't know where I could find you a suit on such short notice," he reached down to the floor and picked up his trunks and said, "I have my suit!"

At the dam I drove directly to the beach and Mr. Martin and his aide had a leisurely swim while Harold Nelson somehow talked a woman out of her towel for the guests to use. But they had to share, since it was the *only* towel on the beach that day. Meanwhile the rest of the group were enjoying their lunch with Ray and Maxine Seely and some of the other ladies at the dam who so generously responded to my cry for help. After the swim Mr. Martin's aide told me, "Now we will

Canadian Secretary of State for External Affairs, Mr. Paul Martin, flanked by the author and by Mr. Arthur Laing, Canadian Minister of Northern Affairs near the bust of President Franklin D. Roosevelt, for whom the lake above Grand Coulee Dam was named. August 5, 1963.

have a relaxed minister." Also he told me that Mr. Martin had been a polio victim and made it a point to swim *every day*. The Canadians were very good sports and we enjoyed having them "test the waters."

The treaty and protocol between the neighbors was signed on September 16, 1964 at the Blaine, Washington Peace Arch. At that time the utilities under the identity of the Columbia Storage Power Exchange made a lump sum payment of $253,900,000 to Canada for its half of the increased power (for 30 years) generated in the U.S. because of the Canadian storage. And the United States agreed to pay $64,400,000 to Canada for the benefits in the U.S. that would accrue here in flood control as a result of the building of Duncan, Arrow and Mica Dams, with combined useable storage of 15,500,000 acre feet of water.

SIZING STUDIES

But going back a bit—at a meeting in Boise, Idaho on September 26, 1961, Bureau representatives from the Commissioner, Chief Engineer, Regional Director and Columbia Basin Project had met to set the parameters for the studies for a third power plant at the dam and on the basis of the 1954 report, had settled on the studies to consider plants of three sizes; 400, 800 and 1,200 megawatts—the latter as the ultimate installation. To that decision I asked, "Why think so small?"

Within minutes, a "veil was lifted" and the concepts were changed—the studies were to proceed with plant capacities of 1.2, 2.4 and 3.6 million kilowatts. At that time, Mr. John Mueller, of the Commissioner's staff in Washington, commented that with an addition of 3.6 million kw at the dam, the load factor would be about that then existing at Boulder Canyon (Hoover Dam)—an installation conceived in the 1920s.

The Bureau finished its plan in January 1965 and the Commissioner forwarded it on February 8 to the Secretary of the Interior, who approved and adopted it only two days later; February 10, 1965. It provided for 12 units of 300,000 kw each. While those studies were ongoing, the staff of the Bureau's Chief Engineer was also reconsidering the size of the units. I became convinced that larger units would make a better fit, and that the ultimate plant should be larger. My mention of such ideas was answered with suggestions that if greater capacity were to be installed it should be in the future utilizing tunnels.

On August 19, 1965 in Wenatchee, Washington, Mr. Charles Luce, Bonneville Power Administrator, predicted that a fourth power house would ultimately be built at the dam to augment the 3.6 million kw development then in the authorization stage. He stated also that the Bureau engineers were studying designs which might result in units rated at 500,000 kw.

On August 26, 1965 I was in the Commissioner's Office in Washington on another matter when Ted Mermel, Chief of the Engineering Division in the Commissioner's Office asked my opinion on the size of the units for the Third Power Plant. I expressed the view that the units should be of optimum size so long as dependability was not sacrificed.

On April 2, 1966, also in Wenatchee, I expressed my concern to Mr. Morgan Durbrow of the Secretary of Interior's Office (Engineering) stating that the Department and the Bureau were going in with a plan that did not provide for maximum site utilization in the initial undertaking for the Third Power Plant. We discussed the size of units which seemed to be his principal interest in the matter at that time. I got the impression that he was satisfied that the 3.6 million kw authorized was as large as could be justified.

I expressed my concerns as to the Bureau's plan being an impediment to the future enlargement of the plant by letter to the regional director and in discussions with some of the Chief Engineer's staff in Denver. And again, in Boise, while we were considering the proposed extreme drawdown of Lake Roosevelt for construction of the plant. I again questioned the Bureau's plan to build a facility for only 3.6 million kw. At that point, the regional director acknowledged that he had received my letter about the matter—"But we are committed to 3.6 million kw and we are not going to change horses in mid-stream." When the meeting recessed briefly I discussed my concerns and my hope that the plant would not be undersized to the Bonneville Power participants, particularly Mr. Bernard Goldhammer and Henderson McIntyre and asked if B.P.A. had made any studies as to the sizing of the Third Power Plant. They had made none, but they were aware that certain projections had been made for capacity exceeding 3.6 million.

Three weeks later, on July 13 when Brigadier General Peter C. Hyzer and others visited the dam, I queried him as to the Corps of Engineers thinking as to the desirable load factor and hydraulic capacity for Chief Joseph Dam and I was advised that they were thinking of a load factor of 20% with about 500,000 cfs hydraulic capacity. Also mentioned was the fact that the Federal Power Commission had urged installations for Libby Dam with a load factor of only 10%. I was then more convinced that with a planned hydraulic capacity of only 250,000 cfs the Third Power Plant plan was greatly undersized.

In mid-December 1966, I received a copy of Mr. Dubrow's report to the Secretary concerning a proposal to maximize the power potential of the Grand Coulee site.

On January 13, 1967 the Secretary of Interior issued a release outlining the program to maximize the power potential at the dam. The forebay would be constructed to accommodate units having 7,200,000 kw capacity. Public law 89-448, dated June 14, 1966, had authorized construction of the Third Power Plant with a rated capacity of 3,600,000 kw with the Secretary of Interior having authority to vary the size and number of units.

On January 13, 1967 the Secretary advised the Congressional Interior and Insular Affairs Committees of the change in plan from 12 units of 300,000 kw to six units of 600,000 kw and that the forebay would be designed and constructed of sufficient depth and width to accommodate a total capacity of 7,200,000 kw. At a meeting in Spokane, I asked Ted Mermel to tell me how the decisions on the Third Power Plant were made. He replied, "Well, the story will never be told, but the Commissioner (Dominy) had been "reading tea leaves," and when he went over to the Department to meet with the Secretary, he proposed six units of 600,000 each."

CONSTRUCTION BEGINS

To get the work for the Third Power Plant under way the Chief Engineer issued Specifications DC 6535 on May 27, 1967 for Modifications to the Left 230 KV Switchyards to accommodate the transfer of all power from the Right Switchyard. That contract was awarded to Jelco, Gibbons and Reed and work began August 8, 1967. Bids for the initial excavation for the Forebay Dam were received on November 7, 1967 and the contract (under Specifications DC 6590) was awarded to Green Construction Co. The construction of the Third Power Plant and the Forebay Dam under Specifications DC-6790 was awarded to Vinnell-Dravo-Lockheed-Mannix on February 26, 1970

and all work under that contract was completed on December 3, 1975. The Third Powerhouse is 1100 feet long, 230 feet high above the foundations and 121 feet wide. The generators are spaced at intervals of 135 feet. There were many other contracts required over the eight years and four months that elapsed from Jelco's start until VDLM's completion. For comparison, the construction of Grand Coulee Dam from Ryan's first shovelful of earth until MWAK diverted the river and built the foundation and CBI finished the structure, ran just 18 days over eight years.

Equipment specifications were issued and awards made for three turbines (specifications DS-6607) with capacity of 820,000 hp, at 285 feet head and 72 rpm, and three generators (specifications DS 6608) of 600,000 kw at 72 rpm with 18,000 volt windings. The manufacturers, Bingham-Willamet and Westinghouse, were to furnish and install the units.

When proposals were offered in response to the invitation (specifications DS-7001) the successful bidder, Canadian General Electric Co., Ltd., proposed units of 700,000 kw capacity, 15,000 volt windings and a speed of 85.7 rpm, the turbines to be of 960,000 hp capacity would be manufactured by the Allis Chalmers Co. The increased generator rating was accomplished by increasing the speed of the units. That combined bid was accepted for $57,829,398 for three units. At time of authorization, the estimated cost of the Third Power Plant was $390,000,000 and the 1985 estimated cost is $761,000,000 with $668,000,000 then expended according to Bureau documents. That gives us an indication of the influence of inflation on long-term projects. Based on 1966 producer prices, the 1984 dollar had shrunk to only $.339.

The Third Power Plant is now in service with the six units producing power and the forebay awaits with ample capacity the future doubling of the plant installation. Mid-1970 estimates indicated the need for additional generation at the dam may occur as early as 1992. Of course, there are problems of debugging these units of unprecedented size, but that has always been the case when "new frontiers" were crossed.

I had no responsibility for the construction of the Third Power Plant, but it has interested me greatly. It is huge by any standards and was a very complex undertaking in that it necessitated removing the right switch yard and about 260 feet of the abutment of the dam, the relocation of the transformer circuits from the Right Power Plant using oil-filled cables through the dam and the complete rebuilding of the left switchyards to accommodate both the Left and Right Power Plants. *And*, it was accomplished while those two power plants continued in service. I shared the responsibility for the operation, maintenance and safety of those operating facilities.

My discussion of the pangs of decision on the unit size and capacity of the Third Power Plant is intended for the pragmatic reader, in the event he or she might wonder about the course of the minds that influenced those elements of the installation. I have referred to my personal files for refreshing my memory on this topic.

In musing over the course of decisions on the Third Power Plant capacity and unit sizes I am impressed by the force of *political* influences on matters that are really first in the realm of *engineering* and *economics*. Too, there was then a great tendency on the part of some of the participants to have their minds set on the past, . "the old-fashioned way," so to speak.

In that connection, though, I should mention something that Dr. John L. Savage stated to me years ago when we were discussing some of the design problems of facilities for the dam. "Engineers," he said, " are too willing to accept their first solution to complex problems and I have difficulty getting them to seriously seek alternatives."

Construction of the forebay for the Third Power Plant required removal of 260 feet of the east abutment section of Grand Coulee Dam. Here one of those blocks is being toppled by blasting. The cofferdam at the left excluded the reservoir water from the work area during construction. February 28, 1969.

CHAPTER XIV
PROGRESS WITH MACHINES OF POWER
Supplement by Raymond K. Seely

RAYMOND K. SEELY

I am pleased to include herein Mr. Raymond K. Seely's comments on the topics which follow in this chapter. Mr. Seely was born on April 10, 1910 at Woodburn, Oregon and graduated from Washington State University with a degree of Bachelor of Science in Electrical Engineering in 1931. He was employed by the Bureau of Reclamation in 1936 at Ephrata, Washington, and was transferred to Grand Coulee Dam as part of Bert Hall's inspection team in 1938. He was reassigned to the electrical and mechanical activity at the dam and power plant in 1941. He was in the U.S. Air Force (radar) for four years, 1942-1946, and attended the Army School at Harvard and M.I.T. for advanced studies. He also took various short courses in management to assure himself that his capabilities in management kept pace with his responsibilities.

In 1946 he returned to his former assignment as electrical engineer and in 1954 became the head of the electrical branch of the electrical and mechanical sub-division at the dam. In 1964 Mr. Seely was promoted to head the Coulee Dam Division of the Columbia Basin Project, from which he retired in 1974. In his assignments at the dam and power plants, Mr. Seely has been in intimate, responsible contact with the installations of power generating and transmission facilities and the problems encountered in operating these great facilities. His observations should give the reader a little insight into the power side of the dam.

TRANSMISSION LINES

Prior to the construction of the transmission lines from Hoover Dam to the California coastal cities, transmission lines had been kept relatively short due to corona at high voltage, high resistance losses causing voltage phase shift, and excessive rise or drop in voltage as the loading would dictate. At low power loads, voltages actually rose, and at high loads, excessive voltage drop would occur. When the first

station service generator (10,000 kw capacity) at the dam was put in service in March 1941, the lines were long and at high voltage in preparation for the larger (108,000 kw) generators then being installed.

The "more or less expected" happened. When the lines were energized from the Bonneville end, the voltage rose at the open end at the dam to a level above the generator capacity. When the lines were charged from the dam end, the current required to charge the line which acted as a long capacitor exceeded the capacity of the generator. This problem was solved by charging half the line from the dam and half from the Bonneville with synchronization being made at the Bonneville Midway Substation (nearly 100 miles from the dam). Line relaying was carefully coordinated to assure that the line wasn't connected to the generator at the dam with no load, a condition which "for certain" would have ruptured the windings of the generator. "Relaying" of transmission lines has made some of the greatest strides of any aspect of the modern (1986) electrical system. In the early days at the Grand Coulee plant, simple directional ground relays and over current relays were the standbys. Those were followed closely with phase comparison relays. The first directional ground relays at the dam were furnished with improperly marked current transformers. That resulted in the relay to "see" the fault in the wrong direction from its protected zone. When a fault then occurred, a couple of miles of high voltage transmission lines melted down before the operator could determine the fault was out on the line and not in the station. This experience demonstrated the necessity for completely testing all current transformers and relay polarity marking upon receipt.

CIRCUIT BREAKERS

As more generators were added to the Bonneville transmission system, larger and faster acting circuit breakers were needed to interrupt the potential faults. The first generator circuit breakers at the dam were rated at 8 cycle 2,500 MVA. But larger capacity and faster circuit breakers were required and manufacturers cooperated with new designs which were tested at the Grand Coulee switchyards. That was the only place in the United States capable of making the test with adequate capacity, which could also be varied by switching lines during light load early morning hours when the testing was actually done. Tests were performed there on various designs (actually full-scale prototype

Columbia Basin Project—Left 230 KV Switchyard. General Electric Company oil circuit breaker tests, March 21 to April 6, 1947. This view shows General Electric Company oil circuit breaker temporarily installed, for interrupting-capacity and line-dropping tests. Nameplate rating: 230/196 KV, 3-cycle, 2.5 million KVA. The manufacturer's representatives, test crew and visitors shown in front of the test breaker are:

J.C. Adams	General Electric Company	Schenectady, NY
J.G. McKnight	U.S.B.R.	Coulee Dam, WA
B.V. Hoard	Bonneville Power Admin.	Portland, OR
V. Hansen	Bonneville Power Admin.	Portland, OR
L.J. Matura	General Electric Company	Schenectady, NY
L.W. Bauer	General Electric Company	Seattle, WA
M.J. Lantz	Bonneville Power Admin.	Portland, OR
P.H. Light	General Electric Company	Schenectady, NY
W. Siegland	U.S.B.R.	Denver, CO
J.F. Spease	General Electric Company	Portland, OR
C.L. Killgore	U.S.B.R.	Denver, CO
W.P. Overbeck	Hanford Engr. Works	Richland, WA
H.L. Levinton	Bonneville Power Admin.	Portland, OR
H.A. Carlberg	Hanford Engr. Works	Richland, WA
W.H. Clagett	U.S.B.R.	Coulee Dam, WA
H.D. Moddel	Hanford Engr. Works	Richland, WA
R.H. Maruchi	General Electric Company	Philadelphia, PA
N.G. Holmdahl	U.S.B.R.	Coulee Dam, WA
H. Ashe	Bonneville Power Admin.	Spokane, WA
C.C. Diemond	Bonneville Power Admin.	Portland, OR
E.B. Rietz	General Electric Company	Philadelphia, PA
A.C. Conger	U.S.B.R.	Coulee Dam, WA
F.R. Schlief	U.S.B.R.	Coulee Dam, WA
J.P. Green	U.S.B.R.	Coulee Dam, WA

April 6, 1947

devices) for the General Electric Co., Westinghouse Co., Allis Chalmers Manufacturing Co., Pacific Electric Co., and the Brown Boveri Co. of Switzerland. Some of these circuit breaker devices were successful and served for a period of time, others were impractical and failed, and some failed in repetitive tests requiring too frequent overhaul to be acceptable.

Mr. Seely and his associates designed and changed many bussing (wiring) schemes to cut down on the duty required on the transmission line circuit breakers. Duties of 10,000,000 kva and 3-cycle were obtained by 1952 and perhaps today, 1986, even higher ratings have been achieved. Mr. Seely has been informed the design capacity of the present (purchased in 1970) Italian Magrini circuit breakers is 52,000 amperes at 550 KV with 2-cycle interrupting time with an assumed ½-cycle relay time. He understands that the circuit breakers incorporated a new design which was much more economical, but is apparently too hard to control. "There is always a difference between laboratory and field performance." To which I would like to add Bill Morgan's sage comment, "Theory is subordinate to practice."

For maintenance and performance reasons, these Magrini breakers were (1985) being replaced with Cogenel & Westinghouse SF6 (sulfur hexafloride) breakers with an interrupting rating of 58,000 amperes at 550 kv with an interrupting time of 1 cycle (¹⁄₆₀ of a second).

The changes in the bussing for the switchyards required many odd current transformers of non-standard ratios during the transition period (while the switchyards were being completely rebuilt without interruption of power.) These transformers were designed at the dam by the Bureau engineers and were built there by Bureau electricians. It is of great credit to all of the engineers, technicians, electricians and linemen that no accident or error occurred in the entire transition from the old to the new switchyard.

OIL-FILLED CABLES

The first generators (station service units) at Grand Coulee Dam were 6900 volt, 12,500 kva, a very standard voltage and conservative rating. These generators were, for all practical purposes, trouble free. The only weak link was the oil-filled cable generator leads. Since the cable was essentially a vertical run, it was impossible to keep the oil sealed in the lower terminators. This was initially corrected by automatically adding oil at the upper end. The ultimate correction was to change the type of cable. This experience gave the project their baptism by fire with oil-filled cables. However, when the Third Power House was planned, space and aesthetic requirements and environmental considerations dictated underground cables with tremendous differences in elevation from the powerhouse to the switchyards. Cable creepage and stress at vertical bends became evident early. Some of the cables have failed and two "emergency" overhead lines have been installed. And it is understood the project program calls for eventual replacement of all of the oil-filled cables with overhead lines.

TURBINES

The chief problems with the turbines back in the 1940s were wheel cavitation and load problems with varying tailwater elevation.

GENERATORS

In 1941 a second generator went into service at the dam. It was rated 108,000 kva with a 13.8 kv winding. Being new to the industry, it lasted about one year. Inspection revealed that corona discharge in the winding slots was destroying the hydrocarbon material in the winding. Contrary to previous practice, the coils were painted with a conductive paint and brought in electrical contact with one side of the slot which effectively bled off the static electricity. These were the largest generators at the time and many problems were encountered with the generators and the turbine water wheels or runners. Hydrogen blisters on the main thrust bearings and insufficient oil across the bearings caused the babbit to be wiped from the bearing surface. The starting technique was changed and pressure lubrication of the bearings was installed. Stress corrosion in the heat exchangers was encountered also.

As it became apparent that trouble was going to be ongoing with the large, new machines, the generator neutral transformers were replaced to give faster ground fault protection. This involved a transformer "loaded up" on a resistor bank, from which relay signals were obtained.

TRANSFORMERS

Transformers have changed little over the years. Those at Grand Coulee were unique only in that they were water-cooled. This caused some problems with leaky heat exchangers. Although static oil pressure was greater than the water pressure, with venturi action this was not always the case.

The 115 kv transformer for unit G-9 caused some operating inflexibility, but it was never considered bad enough to warrant making a change to the 230 kv level. In the early years of the transmission system capacitor potential devices were used to obtain the values of the high voltages. Manufacturers claimed high accuracy and phase stability for these devices. This did not prove to be the case with the weather conditions at the dam. Finally, one potential transformer was secured for a reference for adjusting the potential devices. Ultimately, a foreign manufacturer marketed a reasonably priced design and all the potential devices were replaced with transformers.

ELEVATORS

There are now 12 elevators at the dam, and their maintenance has always been an expensive problem. An investigation revealed that the wearing away of the contacts and gibs was being caused by dust from the concrete in the shafts. It was found to be both practical and economical to seal the walls of the shafts with paint to control the dust.

THIRD POWER PLANT

The Third Power Plant has brought new, larger units which undoubtedly will be beset with many problems for years to come. Those units too, are ahead of the field!

CREDITS
By Ray Seely

Mr. Downs has graciously allowed me to make special mention in this publication of several persons who assisted me in the various "updating efforts" in modernizing the mechanical and electrical facilities at the dam. There were several:

Alvin Conger and John Gregg were very capable help in the Switch Yard changes.

John Forsythe and Robert Small were key men in the modifications in the Power Houses and they were followed by Glenn Barker. Glenn later co-ordinated the job of "juggling" the Switch Yards to make room for the Third Power Plant. His work in this area was almost invaluable to the Bureau and to the contractor.

During the construction and modifications of the Grand Coulee Dam complex, 100 young engineers worked with us for varying lengths of time. Many with their excellent early training have gone on to very satisfactory careers. We are grateful to all of them for their contribution to such a great project.

CHAPTER XV
COSTS AND BENEFITS

THE NAY-SAYERS

Various elements of the population and the media have opposed or spoken in ill or derogatory terms of Grand Coulee Dam ever since it was first envisioned. First there was the proposition that the "coyotes and jackrabbits couldn't use the electricity!" And there were expressions of "judgment" too.

The Honorable Francis D. Cork announced on July 30, 1935, "I now pass to the Proposition of the Grand Coulee, which, in my judgment, is the most Colossal Flop in the History of America."

And *Business Week* October 30, 1938 stated, "Grand Coulee is typical of those things which won't have more use than the pyramid of Egypt."

THE OPTIMISTS

Now again there are those who are seeking to buy the dam as a good investment.

An article in the March 1984 issue of the *Smithsonian Magazine* discussed the anticipated life of modern man-made structures. It predicted that Grand Coulee Dam will be here when all the rest of our modern works are no more. But, it cautioned that, "The dam has one hazard ahead—the next Ice Age. But that isn't expected before the year 6000." Anthropologists of the future should be put on notice that the inspectors did not build the dam, they merely assured that the designs and specifications were actually achieved, and in so-doing, they put their stamp of approval on every bit of it!

To the above observations I urge a bit of caution. The concrete in the dam is of very high quality and in the mass it should not deteriorate. But this is a semi-severe winter climate and the freezing and thawing cycles are enemies of concrete durability. So, one can now see after 50 years or less, that slight surface weathering of the exposed concrete is occurring and will have to be battled from now on. The surface weathering showing on the curbs, sidewalks, roadways, trashrack structures and downstream face of the dam and on the spillway bridge all will require restoratives periodically in future years.

THE INVESTMENT AND THE RETURN FROM SALE OF POWER

The stated cost for construction of the dam and reservoir in 1956 was $161,078,179 and for the power facilities $107,505,605. Minor work—facilities in service (additions) since then add $1,484,470, for a total of $270,068,524. As of September 30, 1985, the cost of the Third Power Plant facilities was $670,854,042. Typical annual costs for operation and maintenance of the dam, reservoir and power facilities were $1,791,000 in 1956 and $2,695,000 in 1966 before the Third Power Plant was started.

The total power generated at the dam to the end of December 1992 was 750,838,784,250 kilowatt hours, of which the Third Power Plant produced 168,634,007,250 kwh. The maximum generation of power (peak load) to date from the Grand Coulee Dam facilities was reached on February 1, 1985 at 5,647,000 kilowatts.

The average rate at which Bonneville Power Administration—the marketing agency for Bureau-produced power in the Northwest—sold power to its customers has varied appreciably over the past 45 years. The average rate received by B.P.A. in the fiscal year 1992 was $0.02148 per kilowatt hour. All amounts shown here are from Bureau records.

THE DAM AND THE ATOMIC ENERGY COMMISSION

The release of water from the Reservoir has had a moderating effect on the temperature of the water in the river below the dam. During construction of the dam, the Columbia River was frozen over for the last three weeks of February 1936 and during the entire month of January 1937. But I think it unlikely that it will ever be frozen over again at that location. The temperature of the water in the reservoir at depth when the reservoir was first filled, approached that at maximum density 39 °F. When flow through the penstocks increased as more units were installed, there was more mixing of the water in the reservoir and the minimum tended to be a bit higher than in earlier years.

The Columbia River is used as the source of cooling water at the plants of the Atomic Energy Commission at the Hanford Works and the temperature of the water has a direct relationship to the production capability there. I recall the discussions with officials of the A.E.C. in our office in Ephrata sometime in the mid-1950s when an agreement was negotiated whereby the Bureau would use the outlet tubes at elevation 1036 in the dam to spill water for the purpose of reducing the temperature of the river water at the Hanford Plants. The A.E.C. then reimbursed the Bureau for its added costs while using the outlet gates instead of just passing the water over the spillway of the dam. This arrangement was brought into play in mid-summer when the river was at its seasonal high temperature with declining discharge.

While discussing the effect of the dams and reservoirs on the river temperature, one of the A.E.C. contractor's (General Electric Co.) representative remarked, "Well, if none of the dams were present we would be better off!" I responded, "Well, if Grand Coulee Dam wasn't there, you wouldn't be here!" It is well known that the availability of power from the dam permitted the construction of the Hanford Works.

As the meeting was closing I remarked, "I suppose you are drawing water for your cooling from the main channel of the river and not some hot side channel slough?"

There was a sudden, rapid exchange of startled glances among our visitors. I happened to see Spud O'Donnell a couple of months later—he had been diving for the A.E.C. at the Hanford Works "Putting in some great, big pipes out in the river!" Too often the obvious is not apparent!

LOCAL IMPACTS

These few photographs which follow are representative of some of the impacts of Grand Coulee Dam on the agricultural, commercial, residential and recreational pursuits and scenic values of the region.

Kettle Falls on the Columbia River with survey party using plane table in the foreground. This feature has been inundated by Lake Franklin D. Roosevelt. It would not have survived long had Grand Coulee Dam not been constructed, because a private utility had applied for a license to construct a dam at this location. August 18, 1937.

Porcupine Bay Campground, one of the many improvements along Lake Roosevelt by the National Park Service for the enjoyment of the recreational oriented citizens. This site is on the south shore of the Spokane River arm of Lake Roosevelt. The lake has numerous excellent beaches for swimming and picnicking. Also, numerous campgrounds have been provided for overnight camping. July 3, 1970.

Steamboat Rock State Park adjacent "Steamboat Rock" (almost an island in Banks Lake) is one of the popular water-based recreational areas of the Columbia Basin Project. This area is administered by the Washington Departments of Game and Parks and Recreation. September 29, 1983.

Summer Falls Power Plant on the Main Canal of the Columbia Basin Project. This beautiful waterfall and adjacent power plant are another benefit of the Grand Coulee Dam. The water flowing here is enroute from Banks Lake to the irrigation of lands in the Project. July 24, 1985.

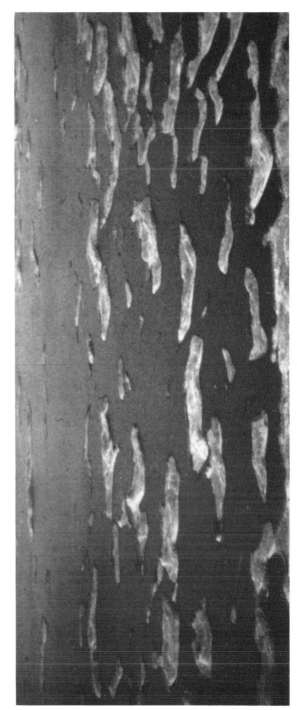

Water stored in Potholes Reservoir behind O'Sullivan Dam made a great, sandy beached land-and-water-scape where the water partially inundated the sand dune area near the mid-section of the Columbia Basin Project. Another great public recreation area administered jointly by the Washington State Departments of Game and Parks and Recreation.

Burke Lake, one of the many seep lakes of the Columbia Basin Project is a highly productive water for Rainbow trout. This opening day crowd braving the cold attest to its popularity. April 19, 1970.

Opening day crowd of fishermen at Susan Lake, another of the many seep lakes of the Columbia Basin Project. Fishing and other water-based recreation are added benefits from the construction of Grand Coulee Dam and the Columbia Basin Project. April 22, 1979.

Satellite scan photo of Columbia Basin Project and surrounding area, showing intensified agricultural development—a result and benefit of Columbia River water from Grand Coulee Dam. Scale 1″ = 8 miles.

Photo by NASA and EROS courtesy of U.S.B.R. taken 570 miles above the earth. August 1981.

OTHER BENEFITS

With the water diverted at Grand Coulee Dam the irrigated lands in the Columbia Basin Project have made a great and beneficial impact on the way of life and the economy of the region. The development of the irrigated farms and the processing of the crops have been the principal elements in the local economy and population growth. Sixty different crops are grown on the project lands with alfalfa hay, potatoes and apples now comprising over one half of the total crop value. Seed crops and fruit are very important crops, with fruit now valued at one sixth of the total gross crop from the irrigated lands of the project. Recently industry related to low-cost power has begun to be an important element of the local economy. This tabulation summarizes the local economic impact from project irrigation and related development.

YEAR	Acres[1] Irrigated	Project[5] Farm Population	Total Project Population	Crop Value $ Thousands	Assessed[2] Value $ Thousands	Project[3] Recreational Use
1930	0	0	9298	0	?	?
1950	4102	140	29,562	503	?	?
1960	289,581	8,414	60,282	39,394	209,900	?
1970	460,641	10,272	63,959	101,710	323,010	1745
1980	500,180	10,199	73,024	275,059[4]	1,170,000	1716[4]
1985	526,751	10,835	77,943	344,000	1,634,000	2138
1990	529,685	12,278		497,995		3359
1991	526,602	12,235		547,734		
TOTAL TO DATE				7,623,248		

1 Includes small acreage served by pumping near Pasco.

2 100% of market value—Grant Co. assessor records only.

3 In thousand visitor days.

4 Adversely impacted by eruption of Mt. St. Helens, May 18, 1980.

5 Full time farm units only.

In addition to the return from the irrigation and the sale of Grand Coulee Power there are other substantial benefits. The reservoir operation provides substantial benefits for flood control each spring as the reservoir is drawn down for storage of the annual flood runoff. The low cost power at the dam is used to pump the water from the reservoir to serve the irrigated lands of the Columbia Basin Project. Also that low-cost pumped water greatly increased recreational prospects, and fish and wildlife benefits accrued to the general public. Revenues from the sale of power produced at the dam are scheduled to repay much of the cost of the irrigation works for which the landowners of the project are the prime beneficiaries.

At present the irrigation of Project lands has only reached the half-way point for the 1,095,000 acres included in the Congressional authorization. Presently the continuing development for service to new lands is at a hiatus. But the Bureau has studies underway to mold the plan to the realities of the 1990s and beyond.

In 1955 the American Society of Civil Engineers awarded the Grand Coulee Dam and Columbia Basin Project the distinction as being one of the seven modern wonders of Civil Engineering in the United States. The entire undertaking is a great achievement in natural Resource Development with its multifold purposes and benefits—supplying water for power, domestic, industrial and agricultural purposes; huge quantities of dependable, clean, low-cost power; flood control to reduce the hazard of devastating floods; 133,057 additional acres of clear, fresh-water lakes; 112,423 acres of Federal lands open to the public; and greatly enhanced fish and wildlife resources for the pleasure and recreational enjoyment of the growing population. Truly the Grand Coulee Dam as a key facility of the project has brought forth "a new star in the flag!"

CHAPTER XVI
AFTER FIFTY YEARS

THE LOW DAM/HIGH DAM QUESTION

Controversy over the use of the waters of our rivers for power and irrigation has in recent years been the subject of much second-guessing by those favoring agendas differing extensively from the concepts that gained wide support for the great water resource developments of the first half of the twentieth century. At least one book has inferred that the U.S. Bureau of Reclamation misled or deceived the Congress and the public when it undertook to build a low dam at Grand Coulee and then changed to the High Dam. My recollection of the events justifying a low dam was that in essence it was a response to a question—"What can you build with $60,000,000?"—of Public Works Administration Funds. It was an effort to put people back to work—AND it did!—at the dam and in the lumber and cement industries of the West and in the steel and manufacturing industries across the land.

On August 14, 1934, the day work started on the MWAK contract to build the Low Dam, Secretary of the Interior Harold Ickes admitted that he had changed his mind and that the Grand Coulee High Dam "Should and must be built." On May 25 and 28, 1935, Dr. Elwood Mead, Commissioner of the Bureau of Reclamation, testified in hearings before the House Committee on Flood Control on H.R. 8057 regarding the status of the Low Dam vs. the High Dam at Grand Coulee. By act of Congress August 5, 1935, the work on Grand Coulee Dam was ratified. Later, on November 28, 1935, the first concrete was placed on the granite bedrock foundation of the Grand Coulee High Dam.

THE FIRST 50 YEARS

Now 50 years have elapsed since the reservoir was filled to its maximum operating level and water was first passed over the spillway. A brief review of how the dam fares might be worthwhile. The weather—hot and cold—the blazing sun, rain and snow, and the mighty Columbia River are persistent adversaries. Too, there are many small to large piping installations of ferrous materials that extend from a direct or indirect connection between the water in the reservoir or tailwater and the galleries within the dam and the power and pumping plants. Valves and bolted covers are used to control and to prevent water from escaping. The trashrack structures protecting the outlet conduits and the power and pumping plant installations have attachments in addition to the removable racks that are of ferrous materials. All are exposed to water and the accompanying tendency to rust and thus decay. Much of that metalwork is located too far below the reservoir surface to permit repair or replacement by divers. And many of the pipe systems connected to the reservoir are too small in diameter to permit access by workmen.

PREVENTATIVE MAINTENANCE

To learn the current practices in preventative maintenance at the Grand Coulee Dam Project, I returned there to get the facts as to the health of this great structure. The management and staff are dedicated to keeping it well maintained, serviceable and operating. Many of the engineers, technicians and craftsmen matured there and know the "what, when, where, why, and how" of the various elements of the Project.

I was brought up to date as to the condition of the facilities and the practices now in vogue to keep it so. Mr. J.S. "Sid" Saunders, Chief of the Engineering and Construction Division, covered the policies and objectives of the ongoing activity. Mr. Gerald Metcalf, Chief of the Construction Branch, Mr. Steve Sauer, Head of the Mechanical Engineering Branch, Mr. H. Michael Lowry, Head of the Geotechnical Section, and Civil Engineer Jan E. Schrader each contributed generously and forthrightly to my understanding of the status of the dam and related features. I was interested also in the results—the payoff (in power delivered)—for the Project and Mr. J.A. Pederson, Chief of the Operations Division, gave me the facts and figures to update those used in the first edition of this book. From him and from Mr. Metcalf, I got a better picture of the scope of the problems and the plans for the upgrading of the big units of the Third Power Plant. As to those units, I will touch only the outline of the undertaking. I greatly appreciate the interest and the assistance of these members of the Project organization.

The need and frequency for routine maintenance and minor and major overhaul of each machine are determined by actual use. Normally, such work is scheduled several years in advance. Replacement parts, including design improvements, are procured as anticipated. In that connection, one of the Government procurement policies—"buy from the lowest bidder"—has resulted in a bit larger variety of inventory of repair parts than would be the case if fewer suppliers had been the vendors for the installations. The maintenance schedule for much of the equipment at the dam, power plants and switchyards must be highly coordinated so that the work can be accomplished without adverse impacts or brownouts in the power grid of the Northwest. With the salability of the hydropower of the Northwest, it is just prudent management or "good business" to have power for sale when there is a demand for it! There exists a very close relationship between the U.S. Bureau of Reclamation, the Bonneville Power Administration, the U.S. Corps of Engineers and numerous other public entities as well as certain private utilities having generation facilities in the Northwest. Power usage is regularly in the low range of the demand cycle during Sundays, holidays and at night. To the extent feasible, routine maintenance is performed during such slack time. Similarly, major maintenance tasks and overhaul of units are scheduled to avoid periods of seasonal high power demand whenever feasible. The machinery of the dam and the plants can and is maintained or replaced almost routinely. But the metalwork embedded in concrete is in a different category. Surfaces exposed to water require protective coatings but the coatings are subject to aging and loss of effectiveness.

PROTECTIVE COATINGS

During the initial construction in the 1930's, coal tar products, applied either hot or cold, were used to protect the metalwork surfaces exposed to water. More specialized coatings were used on the drumgate surfaces exposed to the abrasion of the seals used to confine the water in the drumgate chambers. Of course, with the greatly increased hydraulic capacity of the plants in the present installation, the necessity to use the drumgates for spilling water for flood control operations has been greatly reduced. In fact, no water was spilled for flood control

operations at Grand Coulee from mid-July 1983 to mid-June 1990. The fact remains, however, that the metalwork other than electrical machinery is exposed to water or moisture and so requires suitable durable coatings to overcome or prevent rust and damage. Coal tar products are being replaced by coatings that indicate more efficient protection. Although this trend is just beginning, when coal tar coatings now deteriorate to the point where removal and recoating is necessary, epoxy-based coating materials are used.

EXPOSED CONCRETE

Basic to the longevity of Grand Coulee Dam is the avoidance of practices that would be detrimental to the structure. Those who have had their eyes open have been aware of the accelerated deterioration of concrete walks, curbs and gutters on the streets of our urban areas where various salt, fertilizer and other chemicals have been used to remove accumulations of snow and ice. Such additions do enhance the safety of the public against falls but at great cost. Fortunately, the use of chemicals to control ice on walks and roadways is taboo at the dam and the associated plants. Sand is used to enhance traction on these surfaces—and sand is in more than adequate supply. Some seepage from precipitation has been a minor problem at contraction and/or construction joints in the concrete near the juncture of the dam and the pumping plant. There is no escaping the relentless effect of weather on the exposed surfaces which fortunately are of high quality concrete and well set to resist nature's effort to wear it away. After 50 years, no repair of such surfaces has been necessary. Eventually, all should expect some face lifting of these surfaces. In addition, the freezing and thawing of ice on the spillway concrete has caused some minor scaling of the surface. That, too, may warrant restoration at some time in the future. But 10,000 psi concrete erodes slowly, and cracking has been minimal. The concrete in the dam is like rare old wine—it improves with age.

Within the structure, the floors, walls and ceilings of the galleries and adits are dry and the temperature is comfortably cool except where power cables transit the dam. The gutters in the galleries are of more than adequate capacity so the drainage water coming into the gutters is confined. Water-borne organisms are not of consequence.

While musing about the durability of the concrete surfaces, I recalled that back in 1939 a field experiment was made using an absorptive lining on the form for two successive pours at the downstream face of block 9 at about elevation 1200. When seen over the railing of the dam, this area on the face of the dam has a distinctive character. From a distance of over 100 feet, it appears that the surface is more dense than the abutting surface. The test was made in an effort to find an inexpensive method of removing some of the surplus water from the concrete after placement. That test is out in plain sight for the future to gage.

Under the section headed "Concrete—Damage and Repair" (page 113), I describe the cracking of the concrete in some of the blocks on the downstream face of the dam. During my visits to the dam in the summer of 1992, I was informed that some localized horizontal fracture zones had developed adjacent to the contraction joints in the roadway section of the dam abutments but that no such cracking had been observed in the sidewalk sections alongside the roadway. The concrete in both the roadway and the sidewalks is subject to the same seasonal temperature variations so one would expect that similar "wear and tear" would result. Of course, sand is used to provide traction on snow and ice. No doubt such sand is forced into the cracks by traffic during the spring melt when the contraction joints are at maximum opening. To me this cracking is somewhat of a puzzle.

CONTRACTION JOINTS

The 20-gage non-corrodible metal strips installed across all contraction joints confined the grout from contraction joint grouting. It also successfully prevented the entrance of water from the reservoir or tailwater into the galleries or plants. During the blasting of rock for construction of the Third Power Plant wing dam, water was observed squirting from the contraction joint between blocks 81 and 82. Divers reported that the copper sealing strip in that joint appeared to be damaged. That permitted water from the reservoir to enter the joint. Enough water escaped from the joint at the downstream face of the dam to become a public relations problem. In freezing weather, ice forms where this water emerges and accumulates, creating a bit of a hazard during the spring melt. Thus far, efforts to intercept the flow have met with limited success though perhaps most of the leakage has been diverted into the gallery gutters of the dam.

FOUNDATION DRAINS AND UPLIFT

One of the safeguards at Grand Coulee Dam is the system of drainage holes drilled into the foundation of the dam after the foundation grouting was completed. These 50-foot deep 3-inch diameter holes permit the escape of any water seeping under the dam. This system has been effective as evidenced by observations of the uplift pressures in the rock under the dam. The surface and ground water there contains soluble mineral compounds that tend to precipitate onto passageways, including the drain hole walls. Periodic removal of such accumulations is essential to the proper functioning of drains. A program using ultra-high water pressure and mechanical means is under investigation for possible use. In response to my queries as to the magnitude of the water seeping into the galleries of the dam from the internal tile drains, the contraction joints and the foundation drains, I was assured that the total was not significant and was not monitored. To satisfy my curiosity, I recently called Ray Seely to see what he recalled about the matter. He stated that the total amount of water pumped from the dam drainage system during his administration as head of the Coulee Dam Division was insignificant and never recorded. As an aside—I have been told that proper attire for a visit into the galleries of some dams is "boots and slickers."

SPILLWAY BUCKET

Routine inspection of the spillway bucket and the adjacent river bed is performed in alternate years and after years of major spillway discharge. Results of the recent inspection by the Bureau's SCUBA diving team showed no appreciable change from previous inspections. No deposits were found in the bucket except for minor accumulations of thin "surface scalings" from the face of the spillway. The repaired blocks are in good condition and the prepackt seats used with the caisson are for the most part still intact. Two large furrows of river mantle adjacent to the bucket rise to the approximate elevation of the concrete lip of the bucket. This deposit could be the source of damage to the bucket in the event of an extreme flood when the spillway must be used for major discharge. The prepakt concrete used to fill a pit in the bedrock at the end of the right training wall is reported to be eroding away largely by pothole action. Some of the rubber-lined steel plates in the bucket in blocks 31 and 64 are still intact. Where reinforcing steel is exposed in those locations, there is no apparent effect of damage by electrolysis. The tailwater level at Chief Joseph Dam is now maintained at elevation 960 minimum so there is an added depth to contend with in the diving efforts. That added depth of water, however, should increase the effectiveness of the bucket in the dissipation of energy of the falling water.

OUTLET CONTROL WORKS

With the addition of the greatly increased hydraulic capacity of the Grand Coulee Third Power Plant, the need to retain the lower set of outlet works in the dam was reconsidered and they were removed. Those gate bodies, conduits and drains were then filled with concrete. Also, the operating mechanisms and control piping were removed from the gallery at elevation 950. That gallery as well as that at elevation 1000 then became the routes through the dam for the oil-filled cables carrying power from the Third Power Plant to the 500-KV cable spreading yard on the west side of the river. The outlets at elevations 1036 and 1136 were not affected and remain as facilities available to discharge water from storage for flood control or for emergency use. The gates—80 in number—are continuously exposed to the water in the reservoir. The rubber diaphragms that cause the gates to be tightly sealed deteriorate with age and the constant load from the water seal pressure. Deterioriation of the rubber diaphragms also causes the seal to malfunction when the rubber stiffens or adheres to the adjacent metal frame so that the sealing mechanism fails to close. These sealing materials, both rubber and metals, have undergone several design and specification changes in the last 50 years. Now the metal seal is a new aluminum/bronze product and the quality of the new rubber seal diaphragm is looked upon as an improvement over that obtainable in the past. With 80 gates to keep serviceable, this work is for the most part a continuing, though intermittent, task. The trashracks protecting these conduits do not extend to the reservoir level and in the past divers had to open the hatch covers in these stuctures. That is no longer necessary with the development of the Remotely Operated Vehicle (ROV). The ROV is an underwater mechanical device propelled by water jets and controlled with cable-attached television. At Grand Coulee Dam, diving operations are now limited to depths of 100 feet. Trashracks for the outlet works have not been fully inspected as yet but use of a submersible vehicle

for such service is contemplated. The trashracks for the pump inlets have been inspected using the ROV. All of the trashrack structures for the penstocks and pump inlets will be inspected as a part of the Project structure inspection program. The ROV will be used for these inspections per current plans.

The removal from service and abandonment of the outlet gates at elevation 950, together with the associated piping accessing the reservoir, reduced the risk of failure of such piping systems at some time in the future. The pipes serving the downstream gates also were connected to the surging of the tailwater through the conduit itself. The 6-inch diameter pressure supply pipe and the five similar pipes which serve as drains from the conduit and the gate bodies for each pair of outlet conduits still in use remain as potential sources of "trouble" in the event of undetected pipe decay by rusting. It was reassuring to learn that flexible stainless steel pipe is now available. Pipe systems with high resistance to corrosion are now being used for any replacement of the piping installations directly connected to the reservoir.

While the stainless steel piping will provide vastly increased resistance to corrosion, the critical requirement is the need to have an effective connection or embedment at the face of the dam in the case of those pipes forming the pressure supply system. That MUST prevent water from leaking from the reservoir into the galleries along the exterior surface of the replacement. This seems to be a problem requiring extreme care.

Many years ago, the project experienced at least two episodes of unplanned flows of water into the dam so the need for care is ever present. Fortunately, silt and other sediments in the reservoir no longer affect the operation and maintenance of the dam facilities. Earlier, when slides and bank erosion added much sediment to the water, some note had to be taken of the effect of such abrasives and accumulations on the protective coatings or moving parts of the machinery. Also, it was necessary to avoid the plugging of drains due to the lack of "follow up" in maintenance routines.

RESERVOIR OPERATIONS

Lake Roosevelt is lowered in the spring to provide space behind the dam for the storage of anticipated flood flows which normally occur in May and June. Flood control activity for the entire Columbia River system is in the hands of an inter-agency coordinating committee, with the Corps of Engineers having final authority. Since the completion of the Third Power Plant, the drawdown of Lake Roosevelt is achieved by passing calculated flows through the power facilities except in years of high runoff. In such years, it is necessary to "spill" water to achieve flood control objectives. The lowering of the reservoir for flood control is not "for free" since the loss of head reduces the energy available. But that is the cost of the "insurance" against the anticipated flood damage that would otherwise occur. With current reservoir operations, Lake Roosevelt is near seasonal maximum elevation (1290 feet above sea level) for about 5 months following the June peak. In the beginning, it was full continuously—or nearly so—for several years. Then slides and wave cutting of the shorelands were severe. Now, only the instability of the highway up the San Poil River Valley is in the category of a continuing real problem and wave cutting and slides elsewhere about the shore do not require setting property lines back very frequently. Perhaps it is beginning to live within its bounds.

DRUM GATES

For the viewing pleasure of summer visitors, the drum gates are still lowered for half-hour periods twice daily to spill a 2-inch deep stream of water over the spillway—a spectacular sight. Because the use of drum gates for passing flood flows and spilling for flood control purposes is minimal, the gates are normally held in the fully raised position when practicable. Thus, movement is less frequent and the normal wear and tear on the sealing devices is lessened. In addition, the use of rubber "stops" to supplement the seals has cured the problem of excessive leakage at the pier ends of the downstream seals when the gates are fully raised. The gate seals still require periodic maintenance, though less often.

THIRD POWER PLANT GENERATORS

The manufacture and installation of the large generators at the Third Power Plant, as might have been expected, resulted in certain problems during operation. These units were of greater capacity than

were any others that were in service at the time of design. Additionally, the use of water cooling for the stator windings invited the need for perfect manufacture and installation. Perfection is hard to achieve—and troubles have been experienced as these big units have been put to work. In 1987, a high energy failure of one of the big units occurred with rather spectacular results and fire. The unit was repaired under warranty and returned to service.

In 1991, these six large units generated more than 14 billion kilowatt hours of power. Since these units were designed and installed, major improvements have been made in the technology for large water-cooled hydroelectric generators. To add such improvements and to make the units more reliable, easier to maintain and safer to operate, a contract has been awarded to Siemens Power Corp. The contractor shall design, furnish and install new windings and stator cores in three generators. These changes will increase the operational rating of the generators to 805,000 kilowatts. The replacement schedule for the modified stators, with an outage time of less than 100 days for return to service, calls for the upgrade of one unit per year, with the third unit finished in the fall of 1997.

I and many others had expected that the less than smooth surfaces of the concrete in the Third Power Plant structure walls would result in dust problems during operation and maintenance of the plant. Fortunately that has not occurred and the filtered air system keeps that large building basically dust free.

POWER CIRCUITS

After the disastrous fire destroyed one of the power circuits through the dam, two overhead circuits from the Third Power Plant were constructed across the river. These circuits are not being used but are being held in reserve. The circuits from the Right Power Plant to the 230 KV switchyard were rerouted using overhead lines via block 64 and 84 elevator towers; the circuits from the Left Power Plant to the 230 KV switchyard were rerouted via block 11 and 31 elevator towers. The oil-filled cables through the gallery in the dam at elevation 1200 are no longer used because of the hazard of fire with that installation. The oil-filled cables from the Third Power Plant through the galleries of the dam to the 500 KV cable spreading yard are used with assurance.

Great strides have been made in the automation and remote control features of hydraulic and electrical installations since those first meager steps away from manual control were taken more than 30 years ago. The availability of sophisticated equipment and the willingness to install and use it have increased reliability and improved the response time for the power industry generally. Grand Coulee has been no exception, and the results are considered very satisfactory. I can recall the discussion around the Manager's table of the idea of replacing the manually operated large valve on the penstock drain with a motorized one. Would it be dependable?? Now in 1993, Hungry Horse Dam Power Plant is being monitored from the control center at Grand Coulee Power Plant some 240 miles distant, with the expectation of operating the plant at that distance. Of course sophisticated equipment is expensive and has to be kept in working order. But the system reliability has been improved. When I first discussed the plan to install remote and automatic control systems with the operating personnel at Grand Coulee Power Plant, 122 people were employed in operating the facility. That number was then reduced to 26. Now, with the greatly enlarged facility, there are 77 employees, including 3 apprentices. Additionally, the computer unit has a staff of 18. Time HAS brought change!

PUMPING PLANT DISCHARGE PIPES

From the very beginning of the pumping plant operations, the extent of the vibration of the discharge pipes has been viewed with concern. At that time, each of the discharge pipes was strengthened by adding stiffener rings to decrease the magnitude of the pulsations. But the basic cause of the pulsing was inherent in the "harmonics" of the pump–pipe assembly. Careful observations and tests determined the characteristics of the problem and indicated the changes needed to improve the performance. Scale model tests of the pump–pipe assembly showed that an increase in the number of ports in the pump impeller would greatly improve it. On that basis, a new impeller was procured and installed in unit P-1. It was in service during the 1992 pumping cycle and greatly reduced the pulsing problem—and the concern of the officials. It appears that that problem can and will be licked.

FEEDER CANAL AND PUMP/GENERATORS

The feeder canal extending about 1.6 miles from the outlet of the pumping plant discharge pipes (the headworks) to the Equalizing Reservoir (Banks Lake) has undergone a number of design modifications. When the last six pump units for the Pumping Plant were specified as Pump/ Generator units, a further modification of the canal was made. It was considered necessary to have an open channel structure through the unstable area where the cut and cover portion of the canal had been constructed. No operating or maintenance problems have developed and the operators report no "bore" waves experienced as the result of the use of the generator mode of these units. The units have been used as generators minimally thus far but are in reserve for whenever Banks Lake is near full stage. Pumping for irrigation is "load factored" so that pump power demand is reduced at peak power demands on the Power Grid.

RIVERBANK STABILIZATION

Subsidence of the riverbank slopes at the dam site and vicinity has plagued the construction and operation of Grand Coulee Dam since the outset. After the power plants were put into service, the necessity to control the magnitude of change in the river level became obvious. At that time, certain pumping facilities located in shafts and drifts were installed along the river banks to lower the ground water profile near the river. With the advent of the Third Power Plant and the much wider range of the daily river stage, a much more expanded stabilization program was needed. Extensive deposits of riprap were placed upon the reshaped river bank and an elaborate system for draining and monitoring the ground water level and the stability of the river bank was installed. This system consists of some 673 piezometers, 39 inclinometers, 40 drainage wells and 3 shafts and connected laterals. The installation on the right bank extends for some 6 miles downstream from the dam. The results to date support the belief that the stability thus attained may continue within the current criteria for river stage fluctuations.

PHOTOGRAPHS

Readers of the first edition of this book were invited to let the author know the identity of persons unnamed in the photograph captions. A few did so. Clay Bedford, General Superintendent for CBI, well remembered Felix Kahn of MacDonald and Kahn, Inc. "He wanted to fire me!" My friend Al Johnson identified the concrete crew of which he was a part. Paul Redhead of Eugene, Oregon, brought a box of his grandfather Arthur G. Moulton's working papers, pictures and other records from his days as a vice-president and member of the executive committee of the MWAK Co. Thus, the identity of Mr. Moulton and Col. M.J. Whitson was assured. And, with the assistance of the publisher, ASCE, and the *Pacific Builder and Engineer*, the identity of Leo J. Fischer, Phillip Hart and David Small was revealed. George Brunstead, a Captain for Eastern Airlines, identified himself in the crowd with President Roosevelt as the "three year old in the cap looking at the camera and standing behind my blonde mother Helen." Another youth, Danny Reid, was identified in that crowd by his sister. Mrs. Wm. Krinke identified her husband in the line at the pay window from her cherished copy of that picture. Wayne Lybecker mentioned that the drill foreman was Tony Falbo—Of course! When Tony was ready to blast the bedrock, everyone cleared the site. No one tried to bluff Tony! And Wayne Rawley came by to identify Bill Walker in the photo with Percy Pharr. Park Savage slyly called my attention to the fact that it was Bethlehem Steel Corp. that furnished and erected the trestle—when he recounted his experience checking rivets as it was being built. Later, in reviewing Mr. Moulton's files, I learned that Bethlehem Steel Corp. was in fact a financial backer of the successful bidders. That company built the $1,000,000 trestle on credit with "nothing down and 3 years to pay." It received its payments from CBI as each cubic yard of concrete progressively embedded the trestle.

THE DENVER OFFICE

Luckily, while searching through an old file of records and reports, I found the photograph of the Chief Engineer and his Denver Office staff. The photo was taken on the steps of the New U.S. Customs House in 1931 or 1932 by Mile High Photo Co. which is still in business. In this photo, shown on pages 178–179, each person is identified with a number, correlating to the listing which follows. The 202 individuals shown include all division heads and principal assistants and most of the Denver contingent. I note the absence of Ray Dexter, Arnold Henny and Merle McCleery; perhaps there were others. The Chief Engineer's staff was increased in 1930, 1931 and 1932 and greatly thereafter when the surge of water resource projects got under way. But among those

176

photographed were the ones who conceived and led the design of Owyhee, Hoover and Grand Coulee Dams, all of which were record-breaking structures. It was a GREAT engineering team! Raymond F. Walter (196) was Chief Engineer; Sinclair O. Harper (194) Assistant Chief Engineer; John L. Savage (197) Chief Designing Engineer; William H. Nalder (193) Assistant Chief Designing Engineer; Leslie N. McClellan (198) Chief Electrical Engineer; Byron W. Steele (192) Senior Engineer, Dams; C.M. "Mort" Day (199) Senior Engineer, Mechanical; Ivan E. Houk (201) Senior Engineer, Technical Investigations; E.B. Debler (200) Senior Engineer, Hydrographic; H.R. McBirney (202) Senior Engineer, Canals; H.J. Gault (203) Senior Engineer, All-American Canal; Porter J. Preston (189) Senior Engineer, Colorado River Investigations; Larry R. Smith (190) Chief Clerk; A. Offutt (191) District Counsel.

For this photo, all were grouped by divisions: No. 1–21, Tracing; 22–47, Laboratories; 48–50, Concrete Investigations; 51–70, Mathematical Studies—Arch Dams; 71–72, Research; 73, Field; 74, Hydraulic; 75–78, Engineering Files; 79, Library; 80–81, Legal; 82–103, Clerical; 104–121, Mechanical; 122–126, Canals; 127–154, Electrical; 155, Specifications; 156–188, Dams; 189–203, "Chiefs" and Department heads.

HOW LONG WILL IT LAST?

Before Grand Coulee Dam had received its first cubic yard of concrete, questions were asked about the prospective life of this massive structure. Would it last forever? When at an informal visiting session that question was posed to Dr. Charles P. Berkey, the eminent consulting geologist at the dam, he said NO. He then went on to explain that in geologic terms time is measured in aeons and that the world's work seems to be to wear the earth surface down to sea level. So, at some time in the distant future, Grand Coulee Dam might no longer be visible. But, he added, in usual terms it had a good chance for a long life. It is still a fair question. Many have given it some thought at all levels of expertise. It now appears that if properly maintained using techniques now at hand it might be basically intact indefinitely. A natural or man-made catastrophe, of course, could put it out of commission. But it is located in an area of relatively low risk from volcanic or severe earthquake destruction. A prolonged cycle of high precipitation and lower temperature in the northern hemisphere, however, could return the great ice sheets of the past. That was the menace of the future in that article in the *Smithsonian* magazine. I think I will leave it at that!

"Ice Dam"

When the overburden upstream from the dam slid into the excavated area in the deep trench in the bedrock (block 63−65), excavation was halted until the sliding mass could be restrained. That was done by freezing the toe of the sliding clay mass. Pipes 3 inches in diameter and up to 40 feet in length were installed in holes drilled in the clay in a pattern of an arch some 12 feet in thickness by 120 feet in length. An "ice plant"—ammonia compressors, coolers and condensers—was assembled and the cold brine was circulated through the pipe system. This procedure formed a frozen earth arch in the mass. It provided sufficient support to restrain the sliding material and permit the excavation to be completed and the concrete to be placed "in the dry." Freezing of the mass continued from inception in August 1936 for over one year. The operation was complete after the grouting of the "C" holes located in the upstream toe of the dam was finished.

Bureau of Reclamation—Denver Office

Photo and permission by Mile High Photo Co., Denver

INDEX OF DENVER OFFICE EMPLOYEES

GLOSSARY

ACRE FOOT — Volume of water required to cover one acre to a depth of one foot. Also equal to 325,850 gallons.

ALIDADE — A telescope (with range cross hairs or stadia hairs) mounted on a horizontal axis supported on a metal plate both edges of which are parallel to the telescope. The entire instrument is moved about the plane table and sighted as desired.

BASE LINE — A straight line between two points of a triangulation system, the length of which is accurately determined. It is then used to compute the lengths of all other lines in the triangulation scheme.

BUCKET — The spillway bucket is the formed trough at the bottom of the plunge pool at the downstream limit of the spillway portion of the dam. The bucket is 1650 feet long and shaped to a radius of 50 feet with invert at elevation 870. It terminates at elevation 900 at the apex or lip of the support block, some 20 feet above the bedrock.

BUS — A conductor or group of conductors serving as a connection or group of connections for electrical circuits.

BUS RUNWAY — The inclined structure formed on the face of the dam upon which the bus from the generators to the pumps in the pumping plant was erected.

CALCIUM CARBONATE — A white crystalline compound $CaCO_3$ that forms on the surface of concrete in the hardening process.

CANTILEVER BRIDGE — A bridge whose central span is carried by the projecting ends of two cantilevers the opposite ends of which are anchored to piers.

CAISSON — A watertight structure inside which construction can be performed under water.

CAPACITOR — A Condenser — a device consisting of two or more conductor plates separated from one another by a dielectric and used for receiving and storing an electric charge.

CAPACITOR POTENTIAL DEVICE — An instrument for measuring or controlling differences of electrical potentials.

CLASSIFIER — A machine used to separate sand into a number of sizes.

CLE ELUM DAM — A Bureau of Reclamation dam on the Cle Elum River northwest of Roslyn, Washington.

CURRENT TRANSFORMER — A transformer whose output current is a given ratio to the input current when connected in series to its design load.

C.F.S. — Cubic feet per second.

CHANGE ORDER — A contract document issued to a contractor modifying or altering the requirements and terms of the contract.

COFFERDAM — A structure, usually temporary in nature, constructed in or adjacent water within which the water can be removed by pumping in order to accomplish work in the dry.

COLUMBIA BASIN LAND — The body of lands that could be served with irrigation water from the diversion at Grand Coulee Dam.

CONTRACTION JOINTS — The formed surfaces between the adjacent blocks of concrete in the dam.

CONTROL BAY — The building housing the nerve center of the power plants and the pumping plant. Here are installed indicators—gauges, dials, charts and video screens—and switching and operating devices whereby all important day-to-day operable equipment in the dam, plants and switchyards can be monitored and operated by remote control.

CONTROL CABLE TUNNEL — An extension of the control cable galleries in the dam formed in excavations in the rock or overburden. The control cables from the control bay to the switchyards are housed in these galleries and tunnels.

CONTROL SURVEYS — Surveys establishing accurate horizontal and vertical position locations for reference in construction of the dam and the irrigation system.

COOLING SYSTEM — The means for removing the heat of hydration stored in the concrete as it aged.

CORONA — Corona is a phenomenon surrounding high voltage electric conductors caused by the ionization of the surrounding air. It is manifested by a bluish tint and the discharge is accompanied by a hissing sound and an odor of ozone.

COST PLUS EXTRA WORK ORDER — This is a contract addition for the performance of work where it is not feasible to establish unit prices for the items.

DAM, STRAIGHT GRAVITY — A dam designed and constructed such that the entire water pressure load is resisted by the mass of the structure.

DAM, GRAVITY ARCH — A dam designed and constructed such that some of the water pressure load is transferred to the abutments of the structure by arch thrust.

DIAMOND DRILL — A drill used in boring holes in rock, the tip of which is set with bort diamonds.

DOZER — A bulldozer. Tractor equipped with a horizontal blade attached for moving material by pushing.

DRYDOCK — An enclosure from which water may be pumped out for storage or repair of vessels.

ENERGY DISSIPATORS — Means for converting energy from an active to a passive state.

EQUALIZING RESERVOIR — Banks Lake where the water for irrigation is stored in transit with added capacity to permit pumping and withdrawal at unequal rates.

FACE CAISSON — A watertight structure within which work could be performed below the river level on the face of the spillway of the dam.

FAIL SAFE SYSTEM — A protective system so designed that failure of any of its components will operate as if a failure had occurred in the protected system.

FALSEWORK — Temporary timber framing for support of forms for concrete or other construction.

FAULT — A defect or point of defect in an electric circuit which prevents the current from following the intended route.

FEEDER CANAL — The canal between the pump discharge outlet works and Banks lake.

FISH LADDER — A sloping flume-like structure consisting of a series of steps, over which water flows at moderate velocity to permit fish to ascend an obstacle.

FREEBOARD — Space above the normal high water line to top of the structure.

FRY — Young fish.

FREQUENCY — Number of cycles per second.

GANTRY FRAME — A frame structure raised on side supports so as to span over the work area.

GENERATOR — A machine whereby mechanical energy is changed to electric energy.

GRAND COULEE — An ice age erosional feature of the local terrain, the Grand Coulee extends from Grand Coulee (City) to Soap Lake (City), a distance of 55 miles. The coulee at some points is a thousand feet deep and up to five miles wide.

GROUND FAULT PROTECTION — A means for controlling the flow of ground current so that proper relay performance may be obtained in event of a fault in a prescribed area.

GROUTING — The process of placing or injecting a mixture of cement and water.

GROUTING SYSTEMS — The pipe and outlet system by which grout is delivered into the contraction joints.

HAMMERHEAD CRANE — A symmetrical double cantilevered crane mounted on a gantry at right angles to the direction of travel.

HEAD — The pressure exerted by a body of water.

HEAT EXCHANGER — A system whereby heat is removed or transferred from one body to another.

INTERLOCKS — the flared and jaw-shaped edges whereby adjoining piling are engaged and secured.

IRRIGATION WORKS — Features for diversion and delivery of water for the Columbia Basin Project lands.

KVA — Kilo-volt ampere = 1000 volt amperes. **KW** — Kilowatt = 1000 watts.

LAITANCE — The impurities that rise to the surface of concrete in the hardening process.

LEVEL — An optical instrument used by surveyors in projecting horizontal lines.

LIFT SEAM — A subsurface seam or joint in the bedrock associated with the erosional removal of pressure from overlying deposits.

LOG BOOM — A line of floating, connected timbers used to confine floating materials.

MILLISECOND DELAYS — Blasting detonators callibrated in thousandths of a second.

MVA — Million volt amperes.

NIPPER — A deliveryman.

OIL-FILLED CABLES — Electric cables or conductors surrounded by insulating paper in an oil-filled pipe or conduit.

OUTLET GATE WORKS — The gates and conduits used for control of discharge from reservoirs.

OVERBURDEN — The soil, sand, gravel, etc. overlying the bedrock foundation.

OWYHEE DAM — A large, concrete dam in the Owyhee River in eastern Oregon.

PARADOX GATE — Outlet gate, the design of which provides for seal closure by a wedge on roller tracks that moves the gate body laterally against the sealface during continuing movement of the stem after the gate stops descent.

PENSTOCK — A conduit through the dam for conducting water to the turbine.

PILE BUCKS — Workmen employed on timber piling construction.

PLANE TABLE — An instrument whereby points are located in the field by graphical methods directly onto a map which is fastened to a tripod-mounted table.

POZZOLON — A silicaeous material which in finely divided form and in the presence of moisture will chemically react with calcium hydroxide at ordinary temperatures to form compounds possessing cementitious properties.

POROUS DRAINS — Tile drains cast of porous concrete installed in the dam to intercept seepage, if any, migrating through the concrete of the dam.

PUMP DISCHARGE OUTLET STRUCTURE — The control structure at the outlet of the pump discharge pipes.

PUMP DISCHARGE PIPE — The steel conduit for water from the pumps to the feeder canal.

REFRIGERATION, TON OF — Equals the heat removed from the melting of one ton of ice in 24 hours.

RELAY — A relay is a device by means of which contacts in one circuit are opperated by a change in conditions in the same circuit or in associated circuits.

RESISTOR — A device which introduces resistance into an electric circuit.

RING FOLLOWER GATE — An outlet control gate which in the open position provides a continuous smooth conduit for the passage of water.

ROCK LADDERS — A rock ladder is a vertical "staircase" with short, sloping runs down which the aggregate drops and slides into the pile. Are effective in minimizing breaking of large aggregate when deposited in the stock piles or bins.

SALAMANDER — A fire pot, typically an open-topped, steel barrel used as a source of heat on the work.

SEEP LAKE — Body of water formed in a topographic depression by rising groundwater from irrigation activity.

SHEAR ANGLE — The angle of deflection between the direction of flow and the direction of the log boom.

SIGNAL MAN — Workman directing the positioning of loads suspended from cranes.

SKIP — An open-sided box used for conveyance of materials or men.

SKIPWAY — The track or location upon which a skip is moved.

SPILLWAY — The passage for discharge of water over or around a structure.

STOPLOG — A wooden or steel member used to shut off water in a channel.

STORAGE WORKS — Structures for impounding water for irrigation and other purposes.

STRESS CORROSION — The deterioriation of metals under stress.

SURCHARGE — The controlled storage above the spillway crest.

SWITCHBACKS — Zigzag arrangement of trackage for surmounting steep grades.

SWITCHYARD — The concentration of main electrical connections for generating or transmission systems embodying the generator circuits and the outgoing feeders together with the bus, switching, control equipment and the circuit breakers.

SYNCHRONIZATION — Is achieved in two machines when the voltages are equal and the speeds adjusted so that corresponding instantaneous values of the two waves are reached at the same instant in exact phase.

TIECIRCUIT — Electrical circuit between switchyards.

TRANSIT — A surveying instrument for measuring horizontal and vertical angles.

TRIANGULATION — The operation of measuring the angles in a survey of a portion of the earth's surface in order to determine the triangles controlling the survey.

TRANSFORMER — An apparatus for transferring alternating electric energy from a low to a higher potential or vice versa.

TRANSFORMER CIRCUITS — As used at the dam, transformer circuits extend from the transformers to the switchyards.

TRANSFORMER, GENERATOR NEUTRAL — A protective device (transformer) in series with the generator neutral (to ground) which elevates the grounding current during a fault and quickens the response of the protective relays.

TROLLEY RAIL — An overhead rail upon which a moveable, wheeled carriage is suspended.

TURBINE — A rotary engine actuated by the reaction and/or impulse of a current of fluid subject to pressure.

VOLTAGE PHASE SHIFT — The tendency for phase shift for alternating current on long, high voltage conductors because of the interaction of capacitance and inductance in the associated conductors of the three-phase system.

WELDMENT — A metal part fabricated by welding.

WINDING SLOTS — Recesses formed in the laminated steel within which the coils are installed.

W.P.A. — Works Progress Administration — a governmental agency.

BIBLIOGRAPHY
PART I — OFFICIAL PUBLICATIONS

CONGRESSIONAL DOCUMENTS

House Document No. 103, 73rd Congress, First Session. Columbia River and Minor Tributaries—A general plan for the improvement of the Columbia River and minor tributaries for the purpose of navigation and efficient development of water power, the control of floods, and the needs of irrigation. Two volumes, March 29, 1932.

Report on Columbia Basin Project, by Chief Engineer, U.S. Bureau of Reclamation. January 7, 1932. (Printed in Volume I, Document No. 103. 73rd Congress, First Session, and in Hearings on H.R. 7446. 72nd Congress, First Session.)

Reports of Hearings before Committees on Irrigation and Reclamation in the Senate and House of Representatives:

S-3808. 67th Congress, Fourth Session. Bill authorizing the Secretary of the Interior to investigate and report to Congress upon the Columbia Basin Irrigation Project. December 6, 7 and 13, 1922.

S-2663. 69th Congress, First Session. Bill authorizing the Secretary of the Interior to cooperate with the states of Idaho, Montana, Oregon and Washington in allocation of the water of the Columbia River and its tributaries and for other purposes, and authorizing an appropriation therefor. February 2, 1926.

S-1462. 70th Congress, First Session. Bill for the adoption of the Columbia Basin Reclamation Project and for other purposes. January 11 and 13, 1928.

H.R. 7029. 70th Congress, First Session. Bill for the adoption of the Columbia Basin Reclamation Project, and for other purposes. January 16 and 17, 1928.
Report No. 872 by Congressman Samuel B. Hill of Washington, dated March 10, 1928, to accompany H.R. 7029.

H.R. 7446. 72nd Congress, First Session. Bill to provide for the construction, operation, and maintenance of the Columbia Basin Project in Washington, and for other purposes. May 25, 27, June 1, 2, 3, and 13, 1932.

S-2860. 72nd Congress, First Session. Bill to provide for the construction, operation and maintenance of the Columbia Basin Project in Washington, and for other purposes. June 21, 1932.

CONGRESSIONAL RECORD

Grand Coulee Dam and Irrigation (address on opening bids of Grand Coulee Dam at Spokane, Wash., Dec. 10, 1937). John C. Page. Extension of Remarks by Hon. Chas. H. Leavy, Congr. Record, Dec. 14, 1937, v. 82, no. 24, pp. 2017-2018.

Completion of the Grand Coulee Dam, address on the opening of bids on the dam, Dec. 10, 1937, Congr. Record, Dec. 13, 1937, v. 82, no. 23, p. 1879.

The Grand Coulee Dam, Dec. 10, 1937. Extension of Remarks of Hon. Homer T. Bone, Congr. Record, Dec. 14, 1937, v. 82, no. 24, pp. 2001-2002.

Columbia River and its Resources (from *Portland Oregonian*) Congr. Record, Apr. 1, 1938, v. 83, no. 67, pp. 6036-6037.

Grand Coulee Dam, the World's Mightiest Structure, Congr. Record, Apr. 6, 1938, v. 63, no. 71, pp. 6462-6463.

The struggle for the Grand Coulee Project during the last two years, Congr. Record, June 8, 1938, v. 83, no. 117.

The Columbia Basin's ''Big Four''. Hon. Knute Hill. Article from *Life* Magazine of June 5, 1939, Congressional Record, June 13, 1939, vol. 84, No. 118, pp. 100073-100075.

More About the Grand Coulee Dam and the Columbia Basin Project. Hon. Chas. H. Leavy. Richard L. Neuberger's article in Survey Graphic, July 1939, Congressional Record, July 13, 1939, vol. 84, No. 140, pp. 12701-12704.

U.S. BUREAU OF RECLAMATION

Grand Coulee Dam and Power Plant. Specifications No. 570, pp. 1-89, Completion of Grand Coulee Dam and Left Powerhouse and Foundation for Pumping Plant, Specifications No. 757, pp. 1-161.

Model Tests of Twist Effects in Grand Coulee Dam, Illus., Technical Memo. No. 574, May 31, 1938, 137 pp.

The Grand Coulee Dam and the Columbia Basin Reclamation Project, illus., 48 pages, (small booklet 3" x 6¾", description of construction of dam, etc.) (published by Bureau of Reclamation, Dept. of the Interior). (June 1938).

RECLAMATION ERA

Filming the Columbia Basin Project. C.J. Blanchard. Reclamation Record, July 1923, vo. 14, pp. 236-238.

Columbia Basin Special Commission makes Report. Elwood Mead. Oct. 1925, v. 16, pp. 154-156.

Columbia Basin Project to be Studied. Dec. 1926, v. 17, no. 12, p. 212.

Planning the Columbia Basin Development. Elwood Mead. July 1927, vo. 18, pp. 98-102.

Idaho's Interest in the Columbia Basin Project. T.S. Kerr. May 1929, v. 20, pp. 66-68.

Columbia Basin Project. Roy R. Gill. May 1929, v. 20, pp. 69-70.

Columbia Basin Project Report Shows Feasibility. March 1932, v. 23, pp. 52 and 54.

Columbia Basin Project. F.A. Banks. Jan. 1934, v. 25, no. 1, pp. 12-13.

Columbia Basin Project (address at meeting of National Reclamation Association). F.A. Banks, Jan. 1935, v. 25, no. 1, pp. 12-13.

Details of Cofferdams, Columbia Basin Project. Feb. 1935, v. 25, p. 35.

Grand Coulee Contractors Build Mason City. O.G.F. Markhus. Apr. 1935, v. 25, no. 4, pp. 69-72, 84.

The Belt Conveyor System at Grand Coulee. O.G.F. Markhus. June 1935, v. 25, no. 6, pp. 109-112.

Change of Plan for Grand Coulee Dam. C.H. Carter. July 1935, v. 25, no. 7, p. 135.

Diversion and Care of the River. O.G.F. Markhus. Nov. 1935, v. 25, no. 11, pp. 217-219.

F.A. Banks Discusses Columbia Development. Jan. 1936, vo. 26, no. 1, pp. 3-5.

First Concrete Poured at Grand Coulee Dam. Jan. 1936, v. 26, no. 1, pp. 9-10.

Romance Under the Water. Frank T. Bell. Mar. 1936, vo. 26, no. 3, pp. 62-63.

Schools in the Grand Coulee Dam Area. F.J. Sharkey. April 1936, vo. 26, no. 4, pp. 85, 104.

Grand Coulee Rings Cash Registers. April 1936, v. 26, no. 4, p. 87.

Columbia Basin Investigations and Their Purposes. F.A. Banks. April 1936, v. 26, no. 4, p. 95.

Aggregate Production for Grand Coulee Dam. O.G.F. Markhus. June 1936, v. 26, no. 6, pp. 142-145.

Project Returns Columbia River to its Ancient Course. Aug. 1936, vo. 26, no. 8, pp. 188-189.

Significance of Grand Coulee Dam. F.A. Banks. Dec. 1936, v. 26, no. 12, pp. 278-280.

Facts—Not Fancy. Dec. 1936, v. 26, no. 12, p. 285.

Vista Points on Columbia Basin Project. Dec. 1936, v. 26, no. 12, p. 286.

Grand Coulee Dam, a National Development (by Hon. Harold L. Ickes). Jan. 1938, v. 28, no. 1.

Opening of Bids to Complete Grand Coulee High Dam. Jan. 1938, v. 28, no. 1.

Also One-Third is Ill-Watered. John C. Page. Jan. 1938, v. 28, no. 1.

Reclamation Commissioner Page Recommends Plan for Columbia Basin Development. Jan. 1938, v. 28, no. 1.

Elements of Cost. John C. Page. Feb. 1938, vo. 28, no. 2, p. 21.

The Columbia River Salmon Industry. Ivan Block. Ibid., pp. 26, 27, 28, 29, 30.

Contract Awarded for Completion of Grand Coulee Dam. Ibid., pp. 34-35.

Visitors at Grand Coulee Dam and a Model of the Dam. S.E. Hutton. Ibid., pp. 44, 45, 46, 47.

Columbia Basin Area to Have Net of Roads. Ibid., p. 53.

Giant Dams Compared. July 1938, v. 28, no. 7, p. 129.

What the Grand Coulee Project Means to Washington Agriculture. ibid., p. 146.

Electric House Heating in Mason City. O.G.F. Markhus. Aug. 1938, v. 28, no. 8, pp. 149-150.

Expenditures for Materials—Grand Coulee and Boulder Dam, Aug. 1938, v. 28, no. 8, p. 174.

Visitors at Coulee Dam. Oct. 1938, v. 28, no. 10, p. 207.

Migratory Fish Consultants. Dec. 1938, v. 28, no. 12, pp. 243.

H.A. Parker Now Irrigation Engineer, Columbia Basin Project, Jan. 1939, page 20.

The Grand Coulee of the Columbia. S.E. Hutton. March 1939, page 41.

Coulee Irrigation District Approved, March 1939, p. 46.

Widespread Benefits from Grand Coulee Construction, March 1939, p. 46.

Plans for Control of Migratory Fish at Grand Coulee Dam Approved, inside front cover, April 1939.

Grand Coulee Dam Impounds Rapidly Growing Lake. June 1939, page 130.

Columbia Basin Project (Irrigation features). July 1939, p. 162.

Coulee Dam Construction Makes Steady Progress. July 1939, p. 163.

Concrete Pour Establishes World Record. July 1939, pp. 163, 182.

Harlan H. Barrows Heads Development of Grand Coulee Area. Aug. 1939, p. 207.

The Upper Columbia River Country. Sept. 1939, pp. 230-232.

History of Reclamation in the State of Washington. Oct. 1939, pp. 262-263.

Closure Gates, illus. March 1940, pp. 70-71.

The Snow Lake Tunnel, illus. March 1940, pp. 72-76.

Land Classification — Columbia Basin Project. W.W. Johnston, June 1940, pp. 172-174.

Outlet Works at Grand Coulee Dam, illus. Lloyd V. Froage. August 1940, pp. 215-218.

Joint Investigations — Columbia Basin Project. August 1940, pp. 219-220.

Bonneville to Market Grand Coulee Power. November 1940.

Essential Features of the Bureau's Fish Hatchery at Leavenworth, Washington, illus. — S.E. Hutton. December 1940, pp. 342-345.

Boating on the Grand Coulee Dam Reservoir, illus. December 1940, pp. 349-350.

Migratory Fish Control — Columbia Basin Project. January 1941, vol. 31, no. 1, pp. 1-4.

Beware Land Purchasers. Ibid. pp. 6-7.

Columbia River Reservoir Clearing Project. H.M. Shearer, ibid, pp. 24-25.

Widening Columbia River Channel by Contract. Ibid., p. 7.

Defrosters at Grand Coulee to Combat Ice. Ibid., p. 7.

Penstock Coaster Gates for Grand Coulee Dam. Lewis O. Smith, February 1941, v. 31, no. 2, pp. 38-42.

Concrete for Grand Coulee Dam. Oscar D. Dike. March 1941, v. 31, no. 3, pp. 57-59.

Coulee Dam — The Little Town at the Big Dam. C.E. Benjamin. Ibid., pp. 60-65.

Grand Coulee Dam Generates First Power. April 1941, v. 31, no. 4, pp. 100 and 132.

Sand and Gravel Production for the Grand Coulee Dam. W.T. Mulkay, ibid., pp. 102-103.

Grand Coulee Powerhouse Goes to Work, May 1941, v. 31, no. 5, pp. 133-134.

Special Duties of Concrete Production Department. Oscar D. Dike. June 1941, v. 31, no. 6, pp. 169-171.

Great Power Generators Require Special Shipping Facilities. Ibid., pp. 168 and 171.

Description and Operation of Government Camp. Fred J. Sharkey. August 1941, v. 31, no. 8, pp. 209-211.

Grand Coulee Dam. Big, Strong and Straight. Tom A. Heatfield. September 1941, v. 31, no. 9, pp. 239-241, and 253.

Crops and Livestock a Natural Combination on Northwestern Irrigation Projects, pp. 242-243, and 254.

Cheap Power Brings Industry. Harold L. Ickes. October 1941, v. 31, no. 10, pp. 263-265.

Cement for Grand Coulee Dam. Oscar D. Dike. Ibid, pp. 276-277.

Cooling Grand Coulee Dam Concrete. L.J. Snyder. November 1941, v. 31, no. 11, pp. 285-288.

World's Largest Hydrogenerator Guarded Against Fire. Hu C. Blonk, December 1941, v. 31, no. 12, p. 324.

ENGINEERING EXPERIMENT STATION BULLETINS

Electric House Heating (Experiments in Electric Heating in Mason City, Washington, the Grand Coulee Dam contractor's camp). Homer J. Dana and R.E. Lyle. No. 45, March 1935.

Ibid. No. 46, July 1935.

Ibid. No. 48, August 1935.

Ibid. No. 49, October 1936.

Hydroelectric Power in Washington: a brief on proposed Grand Coulee dams. C.E. Magnusson. University of Washington. no. 78, 1935, pp. 1-29.

HYDROELECTRIC POWER COMMISSION BULLETINS

Modern Builders Surpass Old Egyptians, illus. The Bulletin. Ontario, Canada. December 1939, v. 26, no. 12, pp. 381-393.

U.S. DEPARTMENT OF AGRICULTURE — BUREAU OF CHEMISTRY AND SOILS

Soil Survey of the Quincy Area, Washington. A.W. Mangum and C. Van Duyne of the Department of Agriculture; and O.L. Westover of the Washington Geological Survey. Feb. 8, 1913.

Soil Survey of Franklin County, Washington. Cornelius Van Duyne and J.H. Agee of the Department of Agriculture; and Fred W. Ashton of the Washington Geological Survey. Issued Jan. 13, 1917.

Soil Survey (Reconnaissance) of Columbia Basin Area, Washington. A.T. Strahorn, E.J. Carpenter, W.W. Weir, Scott Ewing and H.H. Skrusekopf of the Department of Agriculture; and A.F. Heck and H.A. Lunt, State College of Washington. No. 28, series 1929.

REPORTS BY THE STATE OF WASHINGTON

The Columbia Basin Irrigation Project. Columbia Basin Survey Commission, Arthur J. Turner, Chief Engineer; J.C. Ralston, Consulting Engineer. 1920.

Columbia River Pumping, Power Project. Willis T. Batcheller, Consulting Engineer, Seattle. February 10, 1922. Not printed. (417 typewritten pages).

Columbia Basin Project. George W. Goethals and Co., Inc. April 7, 1922.

Reports and references by engineers of the Bureau of Reclamation in various issues of the annual reports of the Bureau

References in annual reports of the U.S. Geological Survey.

PART II
TECHNICAL PUBLICATIONS, ENGINEERING AND CONSTRUCTION

AMERICAN CONCRETE INSTITUTE

Aggregate Production for Grand Coulee Dam. G.T. Dodge. Jan. 1936, v. 7, pp. 317-332.

BULLETIN OF THE GEOLOGICAL SOCIETY OF AMERICA

Geology of the rock foundation of Grand Coulee Dam, Washington. William H. Irwin. illus., v. 49, 1938, pp. 1627-1650.

CIVIL ENGINEERING

Columbia River for Irrigation and Power. John S. Butler. Illus., Sept. 1931, v. 1, pp. 1075-1080.

Columbia River—For Irrigation and Power. R.K. Tiffany. Illus., Ibid., pp. 1081-1086.

Improvements of the Columbia River (Grand Coulee site). Col. Thos. M. Robins. Illus., Sept. 1932, v. 2, no. 9, pp. 563-567.

Developing the Columbia River Drainage Basin. Sept. 1934, v. 4, no. 9, pp. 443-459.

Columbia Basin and Grand Coulee Projects. F.A. Banks. Sept. 1934, v. 4, no. 9, pp. 456-459.

Foundation Conditions for Grand Coulee and Bonneville Projects. C.P. Berkey. Feb. 2, 1935, v. 5, no. 2, pp. 67-71.

Construction Plant at Grand Coulee Dam. C.D. Riddle. Oct. 1936, v. 6, no. 10, pp. 639-642.

Experiments Aid in Design at Grand Coulee. Nov. 1936.

Large Plants for Aggregate Production. M.P. Anderson. Illus. (includes Marshall Ford and Grand Coulee Dam). June 1939, v. 9, no. 6, pp. 341-344.

Distinctive Features of Grand Coulee. Jacob E. Warnock. December 1940, pp. 779-782.

Floating Caisson Facilitates Repair of Grand Coulee. L. Vaughn Downs. April 1950, vo. 20, no. 4, pp. 35-39.

THE COLORADO ENGINEER

Placing Concrete at Grand Coulee. J.B. Romans. May 1937, pp. 62-66.

CONSTRUCTION METHODS

Contractors Build Trestle in Fight Against Raging River. Feb. 1936, v. 18, pp. 34-36.

Coulee Dam Concrete Placed from Steel Trestles. March 1936, v. 18, p. 54.

CONTRACTORS AND ENGINEERS MONTHLY

Unique Remedy for Leaning Bridge Pier—Grand Coulee Highway. May 1936, v. 52, no. 5, pp. 1, 37.

Concreting Trestles Another "Biggest Ever" at Grand Coulee Dam. June 1936, v. 32, no. 6, pp. 2, 17.

Facts and Figures about Grand Coulee. June 1936, v. 32, no. 6, p. 4.

A Grouting Tunnel Through Base of Grand Coulee Dam Provides Working Gallery to Check Seepage. July 1936, v. 33, no. 1, pp. 1 and 15.

Greasing the Skids at Grand Coulee. H.W. Young. Jan. 1937, v. 34, no. 1, pp. 18, 41, 49.

Attention to Details Saves Coulee Dollars. H.W. Young. Mar. 1937, v. 34, no. 3, pp. 1, 36, 45.

Giant Cantilever Cranes to Swing Coulee Concrete. illus., Nov. 1938, v. 35, no. 11, p. 7, 17, and 32.

Giant Steel Trestle Erected in 90 days, illus. Dec. 1938, pp. 1, 21, and 32.

"Rubber Railway" at Grand Coulee, Conveyor Belt Shipped in Eight 10-Ton Pieces. Henry W. Young. Illus., Jan 1939, v. 36, no. 1, pp. 1, 11, and 36.

ELECTRICAL WEST

8,000,000 kw. Goal of Columbia River Development; utilization of generating facilities from power, navigation, flood control and irrigation projects. Feb. 1934, v. 72, pp. 16-17.

Grand Coulee; the Key to Columbia River Power. Ibid., pp. 18-19.

8,000,000 kw. is Columbia River Goal; P.W.A. Projects at Grand Coulee and Bonneville. Mar. 17, 1934, v. 73, pp. 409-410.

Wanted: Buyers for One billion Kw.-hr.; Columbia River Development. Feb. 1935, v. 74, pp. 24-32.

Grand Coulee, illustrations. June 1935, v. 74, p. 71.

Grand Coulee Revised Plans Omit Power. July 1935, v. 75, p. 54.

Economic Aspects of Grand Coulee; studies in electric heating and building insulation at Mason City. Oct. 1935, v. 75, pp. 54-55.

Results of Electrical Heating at Mason City. May 1936, v. 76, no. 5, pp. 17-18.

Induction Heating Applied to Coulee Drum Gates. Ralph D. Goodrich Jr., Oct. 1940, v. 85, no. 4, pp. 46-48.

Grand Coulee, Its Story in Pictures. November 1941.

ELECTRICAL WORLD

All Electric City Provides Heating Data, Tables—July 18, 1936, v. 106, no. 29, pp. 35-36 (2185-2186).

THE ENGINEERING JOURNAL

The Columbia Basin Project in the State of Washington, U.S.A. C.E. Webb. Feb. 1936, v. 19, no. 2, pp. 71-74.

ENGINEERING NEWS RECORD

Main Features of Columbia Basin Project. Prof. O.L. Waller. March 4, 1920, v. 84, p. 456.

Report on Columbia Basin Irrigation Project. Elwood Mead. Sept. 24, 1925, v. 95, pp. 502-503.

Columbia Basin Project Reported Feasible. F.E. Schmitt, editor. Illus. June 30, 1932, v. 108, pp. 907-911.

Development Discussed by Engineers. July 14, 1932, p. 20.

Three Big Dam Operations Begin in the Northwest; Construction Operations at Bonneville, Grand Coulee, and Fort Peck. April 5, 1934, v. 112, pp. 441-445. (Grand Coulee pp. 442-444).

Grand Coulee Dam, Washington. Oct. 11, 1934, v. 113, p. 481.

Power, Navigation and Irrigation in Two Projects on the Columbia. Nov. 29, 1934, v. 113, pp. 678-685.

Dam Stresses and Strains Studied by Slice Models. J.L. Savage. Dec. 6, 1934, v. 113, no. 23, pp. 720-723.

Construction Work in Progress. Apr. 18, 1935, p. 574.

Grand Coulee Revised Plans Omit Power. June 13, 1935, v. 114, p. 861.
Editorial Introduction. Grand Coulee Dam issue. Aug. 1, 1935, v. 115, no. 5. (Progress at Grand Coulee). pp. 139-140.

Grand Coulee Project and Dam. Kenneth B. Keener. Ibid., pp. 141-143.

Ten Months' Construction Progress. Ibid., pp. 144-147.

Constructing the First Cofferdam. Ibid., pp. 148-151.

Belt Conveyors for Spoil Transportation: I. Planning for Excavation Disposal. Francis Donaldson. Ibid. pp. 132-153. II. The Belt Conveyors in Operation. pp. 153-155. III. Designing the Belt System. Stanley M. Mercier. pp. 156-157.

Making Aggregate at Grand Coulee. Ibid., pp. 158-160.

Earth Pressure Tilts Pier of Bridge. Nov. 7, 1935, p. 646.

Aggregate Transported Across Columbia by Long Suspension Belt. Nov. 14, 1935, v. 115, pp. 674-675.

Feeder Grizzly on Conveyors. Dec. 12, 1935. p. 821.

Concrete Mixing and Placing at Grand Coulee Dam. Jan. 23, 1936, v. 116, pp. 119-120.

Aggregate Stockpile Tunnels. June 25, 1936, p. 920.

Construction Camp Water Consumption. Sept. 10, 1936, v. 117, p. 364.

Concrete Dispatching System Used at Grand Coulee Dam. Sept. 10, 1936, v. 117, pp. 377-379.

Concrete Bridge Pier Righted After Slide Causes Tilting. Sept. 17, 1936, v. 117, pp. 410-411.

Program for Grand Coulee's Second Cofferdam. Oct. 1, 1936, v. 117, pp. 464-466.

Concrete Improved by Better Control. Oct. 15, 1936, p. 546.

Ice Dam Stops Earth Slide. Feb. 11, 1937, p. 211; Mar. 18, 1937, p. 404.

Cofferdam Leak Checked. Apr. 22, 1937, p. 595.

Contractors Win River Battle. July 1, 1937, p. 13.

New Concrete Placing Record. July 15, 1937, p. 101.

Grand Coulee Cofferdam Removal. Sept. 2, 1937, p. 401.

Grand Coulee High Dam (pp. 1021-1024) and Contract Unit Prices (pp. 20 and 22 of advertising section). Dec. 23, 1937, v. 119.

Wrenching out a Cofferdam. Illus., May 12, 1938, v. 120, no. 19, pp. 677-679. (Grand Coulee Dam).

Steel Gates Close Gaps in Dam. Illus., Dec. 8, 1938, v. 121, pp. 733-734.

Grand Coulee Breaks Concreting Record. June 8, 1939, v. 122, p. 63, (749).

A Low-Cost Concrete Building and Floors (at Coulee Dam). Illus., Aug. 3, 1939, v. 123, pp. 62-63 (156-157).

Concrete Pouring Record Set at Grand Coulee Dam. (397,994 cubic yards in August.) Sept. 28, 1939, v. 123, p. 31 (385).

Making a Water Tunnel Entrance 165 Feet Below lake Surface, Migratory Fish Control. Drawings, March 14, 1940. v. 124, no. 11, p. 53.

Columbia River Bridge Raised 45 Feet. October 10, 1940, v. 123, no. 15, pp. 54-55.

Grand Coulee Dam Nears Completion. Sept. 11, 1941, p. 365.

EXCAVATING ENGINEER

Excavating for Grand Coulee. H.W. Young. Aug. 1936, v. 30, no. 8, pp. 397-401, 432-436.

Freezing Sand at Grand Coulee. Sept. 1936, v. 30, no. 9, p. 476.

World's Biggest. Karl Stoffel. March 1941, pp. 146-149.

ELECTRICAL ENGINEERING

Recent Views of Columbia River Projects; Grand Coulee Dam and Bonneville. Nov. 1938, v. 54, pp. 1272-1273.

The Columbia Basin Project. Alvin F. Darland. Nov. 1937, v. 56, no. 11, pp. 1339-1345.

INTERNATIONAL ENGINEERING

Building Highlights of Grand Coulee. John H.D. Blanks. Illus., Sept. 1938, v. 74, no. 3, pp. 69-74 (con't.)

Building Highlights of Grand Coulee. John H.D. Blanks. Illus., (con't. from Sept. issue), Dec. 1938, v. 74, pp. 197-205.

Clearing Grand Coulee Reservoir Site. November 1941, pp. 152-154.

THE MARQUETTE ENGINEER

The Biggest Thing on Earth. Victor Sumnicht. Nov. 1938, v. 13, no. 1, pp. 15-16.

MECHANICAL ENGINEERING

The Grand Coulee Dam and the Columbia Basin Project. Illus. S.E. Hutton. September 1940, pp. 651-660.

MODERN IRRIGATION

What is the "Scheme of Development Upon Which the Inland Empire is Based?" Fred A. Adams. October 1925, v. 1, pp. 15-16.

World's Largest Irrigation Project Now Being Planned by Four States of Northwest. R.K. Tiffany. Illus., Sept. 1927, v. 3, pp. 16-17, 80.

PACIFIC BUILDER AND ENGINEER

Preliminary work starts on Coulee Dam, portrait of Dr. Mead and tribute to him. Walter A. Averill, ed. Sept. 2, 1933, v. 38, no. 35, p. 27.

Preliminary Data on Design and Construction Features of Coulee Dam. April 14, 1934, v. 39, no. 15, pp. 22-23.

Engineers' Report on Design of Grand Coulee Dam. Ibid., pp. 23 and 30.

Stripping of Overburden at Coulee Dam Six Weeks Ahead of Schedule. Walter A. Averill. Ibid., pp. 24-29.

Slide on Schedule No. 1. I.E. Stevenson. Ibid., p. 29.

The Conference of Consultants on Grand Coulee Dam, at Denver, March 28, 30, 31. James O'Sullivan. Ibid.

Last Minute Dispatch from Grand Coulee Dam. Ibid., p. 31.

Pre-Bid Articles on Grand Coulee Dam. May 5, 1934, v. 40, no. 8, pp. 22-25, 31.

Two Suggested Methods for Diverting the Columbia River at Coulee Dam. James O'Sullivan. June 9, 1934, v. 40, no. 23, pp. 29-30.

What's Doing at Grand Coulee Dam. Ibid., p. 35.

Unit Bids on Grand Coulee Construction Railroad. Ibid., p. 37.

Construction Starts Soon on Grand Coulee Dam. July 7, 1934, v. 40, no. 27, pp. 14, 16, and 27.

Inland Empire Celebrates Opening of Bids on Grand Coulee Project. Ibid., pp. 16-17.

Comparative Unit Bids on Grand Coulee Dam and Power Project. Ibid., pp. 26-27.

The Grand Coulee High Dam . . . And the Irrigation of the Columbia Basin. F.A. Banks. Sept. 1, 1934, v. 40, no. 35, pp. 17-20.

Progress Schedule and Construction Methods for Grand Coulee Dam. Oct. 6, 1934, v. 40, no. 40, pp. 16-18.

Preliminary Construction at Grand Coulee Dam Speeds Ahead. Nov. 3, 1934, v. 40, no. 44, pp. 26-27.

Work Starts on Cofferdams on Coulee Project. Jan. 5, 1935, v. 41, no. 1, pp. 39-40.

Grand Coulee High Dam, With Reclamation and Power Development, Unqualifiedly Endorsed. Ibid., p. 43.

The World's First All-Electric City (Mason City). G.W. Hitchcock. Jan. 19, 1935, v. 41, no. 3, pp. 3-4.

Basis of Findings of Pacific Northwest Regional Planning Commission on the Grand Coulee Dam. March 2, 1935, v. 41, no. 9, pp. 28-29.

M.W.A.K. Rush Grand Coulee Cofferdam. Ibid., pp. 36-38.

M.W.A.K. Completes First Cofferdam at Grand Coulee. Apr. 6, 1935, v. 41, no. 14, pp. 33-34.

How Mason-Walsh-Atkinson-Kier Built World's Largest Cellular Cofferdam in 90 days—Establishing a New Record. May 4, 1935, v. 41, no. 18, pp. 24-31.

All Contractors Hit Ball. Ibid., pp. 42-43.

M.W.A.K. Purchases World's Largest Concrete Plant for Grand Coulee Dam. June 8, 1935, v. 41, no. 23, pp. 26-27.

High Dam at Grand Coulee Assured by Change Order. Walter A. Averill. July 6, 1935, v. 41, no. 27, pp. 30-32.

Building the Construction Highway at Grand Coulee Dam. Sept. 7, 1935, v. 41, no. 36, pp. 34-36.

Moving a Mountain a Mile at Grand Coulee. Oct. 12, 1935, v. 41, no. 41, pp. 28-38.

Producing Aggregate for the World's Largest Concrete Structure. Robert J. Jenks. Dec. 7, 1935, v. 41, no. 49, pp. 26-32.

Manufacturing 4,500,000 Cubic Yards of Concrete for Coulee Dam. Jan. 4, 1936, v. 42, no. 1, pp. 30-36.

Half-Million Yards of Concrete Poured at Grand Coulee. June 6, 1936, v. 42, no. 23, p. 30.

How Pier No. 2, Columbia River Highway Bridge, Was Straightened After It Tipped Out of Alignment. July 4, 1936, v. 42, no. 27, pp. 28-32.

M.W.A.K. Starts Diversion of Columbia. Dec. 5, 1936, v. 42, no. 49, p. 32.

Design and Construction of the Cross-River Cofferdams at Coulee. Donald O. Nelson. March. 5, 1937, v. 43, no. 10, pp. 32-37.

When the Lower Coffer Broke at Coulee. June 5, 1937, v. 43, no. 23, pp. 39-41.

Bids Close December 10th for Completion of Coulee Dam. Nov. 6, 1937, v. 43, no. 45, pp. 30-31, 53.

Contractors Prepare Bids on World's Biggest Concreting Contract. Dec. 4, 1937, v. 43, no. 49, p. 29.

Closure Gates . . . More Pioneering at Coulee. H.W. Young. Dec. 4, 1937, v. 43, no. 49, p. 40.

America's Greatest Builders of Dams Merge to Bid Largest Concreting Contract. Jan. 8, 1938, v. 44, no. 2, pp. 28-29.

Consolidated Builders Tackle Work of Completing Coulee Dam. Apr. 2, 1938, v. 44, no. 14, pp. 38-39.

C.B.I. Wrecks Bridge at Grand Coulee. Apr. 2, 1938, v. 44, no. 14, p. 49.

Refrigerating 11 Million Yards of Concrete. May 7, 1938, v. 44, no. 19, pp. 40-41.

Bethlehem Erects High Trestle; Colby Builds Big Double Cantilever Cranes as C.B.I. Prepares to Pour Concrete. Fred K. Ross. June 4, 1938, v. 44, no. 23, p. 52.

Moving a Mountain of Aggregate at Coulee Dam. Illus. Fred. K. Ross. Sept. 3, 1938, v. 44, no. 36, pp. 51, 53, 58.

New High Trestle Completed on Coulee Dam's 5th Birthday. Illus., Oct. 1, 1938, v. 44, no. 40, pp. 38-40.

Nine Miles of Welding at Coulee Dam. Fred K. Ross. Illus., Nov. 5, 1938, v. 44, pp. 34-36.

Mixing Time Reduced at Grand Coulee. Illus., Dec. 3, 1938, v. 44, pp. 28-29.

H.A. Parker Appointed Irrigation Engineer for Columbia Basin, Portrait, Jan. 7, 1939, v. 45, no. 1, p. 26.

New Set-up in Coulee Graval Pit. Illus., Apr. 1, 1939, v. 45, no. 35, pp. 36-39.

New World's Record Set at Coulee Dam (20,684 Cubic Yards of Concrete in 24 Hours). Illus. June 3, 1939, v. 45, no. 22, pp. 28-33.

20,684 Cubic Yards in 24 Hours, New World's Record at Coulee Dam. Illus., July 8, 1939, v. 45, no. 27, pp. 30-33.

Handling Concrete in the Blocks at Coulee. Illus. Aug 5, 1939, v. 45, no. 31, pp. 25, 65.

Cranes are Vital Cog in Placing Concrete at Coulee. Illus., Sept. 2, 1939, v. 45, no. 35, pp. 36-39.

Human Safety Emphasized at Grand Coulee Dam. Illus. May 1940, pp. 38-42.

How ''Tip'' O'Neal Highballs His Columbia Canyon Job. Illus., May 1940, pp. 28-34.

Pumpcreting at Grand Coulee. Illus. June 1940, pp. 36-38.

Here's What Will be Done at Grand Coulee Dam in 1941. May 1941, pp. 33-34.

PACIFIC ROAD BUILDING AND ENGINEERING REVIEW

Grand Coulee Produces Defense Power. April 1941, pp. 8-13.

POWER

Facts and Figures on Grand Coulee Dam. Illus., Oct. 1938, v. 82, no. 10, pp. 78-80 (550-552).

So Big! Main Gates Weigh 100 Tons, 45 Feet High and 12 Feet Wide, 60 Main Gates and 60 Emergency Gates, 8 ½ Feet in Diameter. View only. Dec. 1938, v. 82, no. 12, p. 128 (706).

POWER PLANT ENGINEERING

Grand Coulee Goes to Work. April 1941, pp. 72-73.

SOUTHWEST BUILDER AND CONTRACTOR

All Houses in Mason City with 3000 Population, Electrically Heated. O.G.F. Markhus. Apr. 26, 1935, v. 85, no. 17, pp. 10-11.

First Million Yards of Concrete Placed in Grand Coulee Dam. Aug. 21, 1936, v. 88, p. 10.

Pour 7,000 Yards of Concrete Daily at Grand Coulee. Oct. 23, 1936.

One-Third of West Ill-Watered, Only Hope of Development in Irrigation (Coulee bid Opening). Dec 17, 1937, v. 90, no. 25, p. 15.

Processing Ten Million Cubic Yards of Sand and Gravel at Grand Coulee. Illus., Jan. 6, 1939, v. 93, pp. 10-11.

Coulee Dam Pump Research Program and Its Significance Explained. J.W. Daily. June 2, 1939, v. 93, no. 22, pp. 18-19.

Distribution of Materials by States, Used in Boulder and Grand Coulee Dam. June 9, 1939, v. 93, no. 23, p. 15.

THE WELDING ENGINEER

Grand Coulee Construction Work Speeded with Use of Welding and Cutting. H.W. Young. June 1936, v. 21, no. 6, pp. 28-30.

WESTERN CONSTRUCTION NEWS

Grand Coulee Dam and Power Plant Specifications. A.G. Darwin. April 1934, v. 9, pp. 103-114.

Bids Opened at Spokane for Grand Coulee Dam and Power Plant. B.E. Bjork. July 1934, v. 9, no. 7, pp. 224-229.

World's Largest Construction Conveyor Speeds Excavation at Grand Coulee. March 1935, v. 10, no. 3, pp. 80-82.

Cofferdams 3,000 feet Long Built at Grand Coulee. June 1935, v. 10, no. 6, 162-164.

Grand Coulee Excavating Program Modified to Meet Changed Designs. Sept. 1935, v. 10, no. 9, p. 53.

Slides in West Abutment Area Are Problem at Grand Coulee Dam. Sept. 1935, v. 10, no. 9, pp. 258-259.

Preparing Millions of Yards of Aggregate for Grand Coulee Dam. Nov. 1935, v. 10, no. 11, pp. 310-315.

Grand Coulee Dam Concreting Plant. Chas. Thompson (M.W.A.K.). Feb. 1936, v. 11, no. 2, pp. 31-36.

Progress Notes From the Grand Coulee Project. March 1936, v. 11, no. 3, p. 89.

Concreting Block 40 at Grand Coulee Dam. March 1936, v. 11, no. 3, pp. 92-94.

Grand Coulee Operations accelerated to Rapid Pace. May 1936, v. 11, pp. 164-165.

Details of Concreting Procedure at Grand Coulee. Sept. 1936, v. 11, no. 7, pp. 209-214.

Frozen Earth Used to Stop Slides at Grand Coulee. Sept. 1936, v. 11, no. 7, pp. 300-301.

Diverting the Columbia at Grand Coulee with Timber Cribs and Gravel Fills. Dec. 1936, v. 11, no. 12, pp. 386-389.

Plans for Completion of the Grand Coulee Dam. Illus. Dec. 1937, v. 12, pp. 479-482.

Work Resumed at Grand Coulee With Reconditioning of Plant. July 1938, v. 13, no. 7, pp. 275-278.

Models Used to Demonstrate Grand Coulee Dam Features. C.E. Benjamin. Feb. 1939, v. 14, pp. 57-59.

Erecting a 3,600-foot Steel Trestle for Placing Grand Coulee Concrete. H.S. Brabrook. Illus., Feb. 1939, v. 14, pp. 43-44.

Concrete Placing at Grand Coulee, From Gravel Pit to Forms. Illus., June 1939, v. 14, no. 6, pp. 198-203.

Removing Cofferdams with Under-Water Torch. February 1940, p. 55.

Crane Servicing Procedure at Grand Coulee Dam. Illus., March 1940, v. 15, no. 3, pp. 94-95.

Solving Slide Control Problems. June 1940, p. 204.

Cement Pumping Helped Set the Concreting Record. August 1940, p. 268.

LE TECHNIQUE DES TRAVEAUX

Le Barrage de Grand-Coulee sur le Fleuve Columbia (Washington, E.-U.) Illus. R.C. Skeret et L. Gain. Apr. 1938, v. 14, no. 4, pp. 206-220.

PART III — NATIONAL MAGAZINES

AMERICAN MAGAZINE

The World's Largest Engineering Wonder. Illus., Richard L. Newberger. Jan. 1938, v. 125, no. 1, pp. 14-15, 134-137.

THE ARGUS

A Hydroelectric Empire. Richard L. Neuberger. Illus. Grand Coulee Dam. Feb. 25, 1939, v. 46, no. 8, Golden Jubilee Number, pp. 21-26.

THE ATLANTIC

Great Dam (Grand Coulee) Map. Stuart Chase. Nov. 1938, v. 162, no. 5, pp. 593-599.

COLLIERS

Power in the Wilderness. Walter Davenport. Sept. 21, 1935, v. 96, pp. 10-11.

FORTUNE

Grand Coulee. July 1937, v. 16, no. 1, pp. 79-89, 148, 150, 153, 159, 160.

LIFE

Roosevelt Builds the Biggest Dam and Envisions a New Society. Oct. 11, 1937, v. 3, no. 15, pp. 34-39.

Irrigation makes the Northwest Land Bloom. Illus. of Grand Coulee and Boulder Dams. June 5, 1939, v. 6, no. 23, pp. 15-23, and 46.

NATION

Grand Coulee. J. Rorty. Mar. 20, 1935, v. 140, pp. 329-331. (Discussion v. 140, pp. 446-448, Apr. 17, 1935; v. 141, pp. 101-102, July 24, 1935).

Miracle in Concrete. Richard L. Neuberger. June 1940.

NEWS WEEK

Grand Coulee Project to Start. July 7, 1934, v. 4, pp. 6-7.

POPULAR MECHANICS MAGAZINE

The largest Thing Ever Build. Illus., Grand Coulee Dam Under Construction. April. 1940, v. 73, no. 4, pp. 546-549, 130-A to 131-A.

Paul Bunyan's Pond. Karl Stoffel. February 1941.

POPULAR SCIENCE

World's Greatest Dam (Grand Coulee) to Create an Electrified Paradise. Walter E. Mair. Feb. 1936, v. 128, no. 2, pp. 11-13, 100.

READERS DIGEST

Great Dam (article on Grand Coulee Dam condensed from Atlantic Monthly). Stuart Chase. Jan. 1939, v. 34, no. 201, pp. 85-'88.

REVIEW OF REVIEWS

More Power for the Northwest: Grand Coulee Project. Jan. 1934, v. 89, pp. 48-49.

White Power for the Northwest; Grand Coulee Project. S. Limerick. Aug. 1934, v. 90, pp. 52-53.

SATURDAY EVENING POST

Eighth Wonder of the World. Robert O. Case. July 13, 1936, v. 208, pp. 23 +

Great Works. Garet Garrett. Illus., April 8, 1939, v. 211, no. 41, pp. 4-6, 86-93. (Describes Grand Coulee and other projects.)

The Great Salmon Mystery. Richard L. Neuberger. September 13, 1941, pp. 20-21, 39-45.

SCIENTIFIC AMERICAN

On a Natural Damsite at Grand Coulee. G. Kirkpatrick. Apr. 1935, v. 152, pp. 198-200.

Grand Coulee Progresses. R.G. Skerrett. Illus., Dec. 1938, v. 159, no. 6, pp. 296-299.

SURVEY GRAPHIC

The Columbia Flows to the Land. R.L. Neuberger. Illus., July 1939, v. 28, no. 7, pp. 440-445 and 451-464.

PART IV — TRADE MAGAZINES

BETTER ROADS

Extensive Road Relocation at Grand Coulee Dam. November 1941.

COMMERCIAL CAR JOURNAL

The Grand Coulee Dam is a Grand Grind for Motor Trucks. J.V. Devine. Jan. 1937, v. 52, no. 5, pp. 26-29, 98, 100.

COMPRESSED AIR MAGAZINE

Developing the Mighty Columbia. C.H. Vivian. Sept. 1935, v. 40, no. 9, pp. 4815-4821.

Grand Coulee Dam. H. O'Connell. Oct. 1935, v. 40, pp. 4840-4844.

Construction Methods at the Grand Coulee Dam. Ibid., pp. 4848-4855.

Housing Grand Coulee Workers. Ibid. pp. 4845-4847.

Compressed Air at Grand Coulee Dam. H.W. Young. Dec. 1936, v. 41, no. 12, pp. 5184-5189.

Uncle Sam to Nurse Forty Million Little Fish. Illus. June 1940, v. 45, no. 6, pp. 6174-6178.

DUPONT MAGAZINE

The Grand Coulee Dam. Hugh D. Lavery. May 1936, v. 30, no. 5, pp. 11-15.

ELECTRICAL DEALER

It's a New ''Home on the Range.'' H.W. Young. June 1936, v. 17, no. 6, pp. 12-14.

ELECTRIC WHOLESALING

Electrical Wholesalers at Grand Coulee. J.B. Canning. Dec. 1937, v. 18, no. 12, pp. 10-11, 26.

THE HIGHWAY MAGAZINE

Today's Colossus, Grand Coulee Dam. Harriet Geithmann. July 1938, pp. 148-151, (other illus. on p. 158.)

INDUSTRIAL NEWS

Grand Coulee, "The Bigging Thing on Earth." Apr. 1938, v. 8, no. 4. (published by Gates Rubber Co., Denver, CO)

THE JEFFERY MANUFACTURING CO.

Pebbles on Parade (aggregate plant). n.p., n.d.

Rivers of Dirt (excavation conveyor described). n.p., n.d.

THE OREGON MOTORIST

Harnessing the Columbia. July 1935, v. 15, no. 5, pp. 10 and 18.

OVERLAND TRAILS

America's Last Frontier. Ashley E. Holden. Dec. 1935, v. 1, no. 1, pp. 15-16.

PAPER TRADE JOURNAL

Power Possibilities in Northwest. C.C. Hockley. Sept. 20, 1934, v. 99, pp. 85-86.

THE RED CROSS COURIER

Safety at Grand Coulee Dam. Howard O. Danford and Franklyn M. Johnson. July 1941, v. 12, no. 1, pp. 3, 12.

ROCK PRODUCTS

Handles 2,500 Tons per Hour: Wastes 1,000 to 1,500 Tons of Sand. Edmund Shaw. March 1936, v. 39, pp. 30-41.

Preparation of High Specification Concrete Sand at the Grand Coulee Dam; (abstract). A. Anable. Ibid., pp. 41-42.

Aggregate Production for Grand Coulee Dam. (abstract). Ibid., pp. 42-43.

SPOKANE CHAMBER OF COMMERCE

The Columbia Basin—Grand Coulee Project. Illus., May 1938, 40 pp.

SHELL PROGRESS

Digging In at Grand Coulee. Aug. 1935, pp. 4-5, 12,

STEEL

Elaborate Conveyor Systems Set Up at Grand Coulee Dam Diggings. A.F. Brosky. July 22, 1935, v. 97, pp. 24-27.

Steel Trestles to be Embedded in Concrete at Grand Coulee Dam. May 11, 1936, v. 98, no. 19, pp. 58-59.

WESTERN RECLAMATION

Water Districts Under Grand Coulee. James O'Sullivan. Sept. 1939, v. 2, no. 9, pp. 10, 14.

THE VALVE WORLD

First Unit of Grand Coulee Dam Completed a Year Ahead of Schedule. F.T. Trenery. Jan.-Feb. 1938, v. 35, no. 1, pp. 22-26.

PART V — BOOKS

Behind the By-Line Hu. Hu Blonk.—Clark—1992, 262 pages (Memories of the Dam newsman Hu Blonk, Wenatchee, Wa)

Our Promised land. Richard L. Neuberger. The Macmillan Company, N.Y., 1938, 398 pages. (Account of Grand Coulee Dam, Columbia Basin Project and the Northwest.)

Concrete Manual. U.S. Dept. of Interior, Bureau of Reclamation.

PART VI — NEWSPAPERS

BARRONS

Grand Coulee—A Giant Power Threat; a description of the Columbia River project and its $63,000,000 first unit. L. Stanley. Dec. 25, 1933, v. 13, pp. 3, 8-9.

CHRISTIAN SCIENCE MONITOR

Grand Coulee, Industrial Kingdom. Richard L. Neuberger. July 5, 1941, pp. 3, 14.

NEW YORK TIMES

Industry Drawn by Cheap Power, Pacific Northwest Believes its Old Dream is Now to be Realized. Richard L. Neuberger. January 7, 1940.

Mightiest Man-Made Thing. Richard L. Neuberger. March 16, 1941, pp. 10-11, 18.

THE OREGONIAN

Cruising for Fun (A Motorlog to Grand Coulee Dam, to be World's Largest Man-Made Object on its Completion). Richard L. Neuberger. July 18, 1937, p. 2 of magazine section.

Sin and Salvation at Grand Coulee. Richard L. Neuberger. July 25, 1937, pp. 1, 8 of magazine section.

Blueprint of an Empire. Richard L. Neuberger. November 30, 1941, pp. 1, 4.

SEATTLE TIMES

Grand Coulee Dam Greatest Structure Built by Man. R.J. Harvison. July 12, 1936, v. 59, no. 194, pp. 12, 15 of magazine section.

SPOKANE DAILY CHRONICLE

Golden Jubilee Edition. May 23, 1936.

SPOKESMAN REVIEW

August 15, 1934

Progress Edition. Jan. 26, 1936.

Progress Edition. Jan. 17, 1937.

Progress Edition. 1941.

WENATCHEE DAILY WORLD. Coulee Dam Edition.

Various

NOTE: This Bibliography was abstracted from the Columbia Basin Project Histories 1932 to 1941 and covers the early planning and the period during which the dam was under construction by MWAK and C.B.I.

Henry J. Kaiser, Edgar Kaiser
Charles A. Shea, Jack McEachern, Harry Morrison
Les Corey, Paul Wattis, Gil Shea
Felix Kahn, Phillip Hart, Tom Price